Environmental Health Criteria 78

DITHIOCARBAMATE PESTICIDES, ETHYLENETHIOUREA, AND PROPYLENETHIOUREA: A GENERAL INTRODUCTION

Published under the joint sponsorship of
the United Nations Environment Programme,
the International Labour Organisation,
and the World Health Organization

World Health Organization
Geneva, 1988

The **International Programme on Chemical Safety (IPCS)** is a joint venture of the United Nations Environment Programme, the International Labour Organisation, and the World Health Organization. The main objective of the IPCS is to carry out and disseminate evaluations of the effects of chemicals on human health and the quality of the environment. Supporting activities include the development of epidemiological, experimental laboratory, and risk-assessment methods that could produce internationally comparable results, and the development of manpower in the field of toxicology. Other activities carried out by the IPCS include the development of know-how for coping with chemical accidents, coordination of laboratory testing and epidemiological studies, and promotion of research on the mechanisms of the biological action of chemicals.

ISBN 92 4 154278 0

The designations employed and the presentation of the material in this publication do not imply the expression of any opinion whatsoever on the part of the Secretariat of the World Health Organization concerning the legal status of any country, territory, city or area or of its authorities, or concerning the delimitation of its frontiers or boundaries.

The mention of specific companies or of certain manufacturers' products does not imply that they are endorsed or recommended by the World Health Organization in preference to others of a similar nature that are not mentioned. Errors and omissions excepted, the names of proprietary products are distinguished by initial capital letters.

ISSN 0250-863X

PRINTED IN FINLAND

87/7745 — VAMMALA — 5500

CONTENTS

WHO TASK GROUP ON ENVIRONMENTAL HEALTH CRITERIA FOR DITHIOCARBAMATE PESTICIDES, ETHYLENETHIOUREA (ETU), AND PROPYLENETHIOUREA (PTU)

Members

Dr U.G. Ahlborg, Unit of Toxicology, National Institute of Environmental Medicine, Stockholm, Sweden (*Vice-Chairman*)

Dr H.H. Dieter, Federal Health Office, Institute for Water, Soil and Air Hygiene, Berlin (West)

Dr R.C. Dougherty, Department of Chemistry, Florida State University, Tallahassee, Florida, USA

Dr A.H. El Sabae, Pesticide Division, Faculty of Agriculture, University of Alexandria, Alexandria, Egypt[a]

Dr A. Furtado Rahde, Ministry of Public Health, Porto Alegre, Brazil (*Chairman*)

Dr S. Gupta, Department of Zoology, Faculty of Basic Sciences, Punjab Agricultural University, Ludhiana, Punjab, India[a]

Dr L.V. Martson, All Union Scientific Research Institute of the Hygiene and Toxicology of Pesticides, Polymers, and Plastics, Kiev, USSR[a]

Dr U.G. Oleru, Department of Community Health, College of Medicine, University of Lagos, Lagos, Nigeria

Dr Shou-Zheng Xue, Toxicology Programme, School of Public Health, Shanghai Medical University, Shanghai, China

Observers

Dr R.F. Hertel, Fraunhöfer Institute for Toxicology and Aerosol Research, Hanover, Federal Republic of Germany

Dr E. Kramer (European Chemical Industry Ecology and Toxicology Centre), Dynamit Nobel A.G., Cologne, Federal Republic of Germany

Mr G. Ozanne (European Chemical Industry Ecology and Toxicology Centre), Rhone Poulenc DSE/TOX, Neuilly-sur-Seine, France

Mr V. Quarg, Federal Ministry for Environment, Nature Conservation and Nuclear Safety, Bonn, Federal Republic of Germany

Dr U. Schlottmann, Chemical Safety, Federal Ministry for Environment, Nature Conservation and Nuclear Safety, Bonn, Federal Republic of Germany

Dr M. Sonneborn, Federal Health Office, Berlin (West)

Dr W. Stöber, Fraunhöfer Institute for Toxicology and Aerosol Research, Hanover, Federal Republic of Germany

---

[a] Invited but unable to attend.

Observers (contd)

Dr D. Streelman (International Group of National Associations of Agrochemical Manufacturers), Agricultural Chemicals Registration and Regulatory Affairs, Rohm & Haas, Philadelphia, Pennsylvania, USA

Secretariat

Mrs B. Bender, International Register for Potentially Toxic Chemicals, Geneva, Switzerland

Dr A. Gilman, Industrial Chemicals and Product Safety Section, Health Protection Branch, Department of National Health and Welfare, Tunney's Pasture, Ottawa, Ontario, Canada (*Temporary Adviser*)

Dr L. Ivanova-Chemishanska, Institute of Hygiene and Occupational Health, Medical Academy, Sofia, Bulgaria (*Temporary Adviser*)

Dr K.W. Jager, International Programme on Chemical Safety, World Health Organization, Geneva, Switzerland (*Secretary*)

Dr E. Johnson, Unit of Analytical Epidemiology, International Agency for Research on Cancer, Lyons, France

Dr G. Rosner, Fraunhöfer Institute for Toxicology and Aerosol Research, Hanover, Federal Republic of Germany (*Temporary Adviser*)

Dr G.J. Van Esch, Bilthoven, Netherlands (*Temporary Adviser*) (*Rapporteur*)

NOTE TO READERS OF THE CRITERIA DOCUMENTS

Every effort has been made to present information in the criteria documents as accurately as possible without unduly delaying their publication. In the interest of all users of the environmental health criteria documents, readers are kindly requested to communicate any errors that may have occurred to the Manager of the International Programme on Chemical Safety, World Health Organization, Geneva, Switzerland, in order that they may be included in corrigenda, which will appear in subsequent volumes.

ENVIRONMENTAL HEALTH CRITERIA FOR DITHIOCARBAMATE PESTICIDES, ETHYLENETHIOUREA (ETU), AND PROPYLENETHIOUREA (PTU)

A WHO Task Group on Environmental Health Criteria for Dithiocarbamate Pesticides, Ethylenethiourea, and Propylenethiourea met at the Fraunhöfer Institute for Toxicology and Aerosol Research, Hanover, Federal Republic of Germany, from 20 to 24 October 1986. Professor W. Stöber opened the meeting and welcomed the members on behalf of the host Institute. Dr U. Schlottmann spoke on behalf of the Federal Government, which sponsored the meeting. Dr K.W. Jager addressed the meeting on behalf of the three co-sponsoring organizations of the IPCS (UNEP/ILO/WHO). The Task Group reviewed and revised the draft criteria document and summarized the health risks of exposure to dithiocarbamate pesticides.

The drafts of this document were prepared by DR L. IVANOVA-CHEMISHANSKA, Institute of Hygiene and Occupational Health, Sofia, Bulgaria, and DR G.J. VAN ESCH, Bilthoven, the Netherlands.

The efforts of all who helped in the preparation and finalization of the document are gratefully acknowledged.

* * *

Partial financial support for the publication of this criteria document was kindly provided by the United States Department of Health and Human Services, through a contract from the National Institute of Environmental Health Sciences, Research Triangle Park, North Carolina, USA - a WHO Collaborating Centre for Environmental Health Effects. The United Kingdom Department of Health and Social Security generously supported the cost of printing.

ABBREVIATIONS

ADI	acceptable daily intake
BSP	sulfobromophthalein
DDC	diethyldithiocarbamate
DIDT[a]	5,6-dihydro-3*H*-imidazo (2,1-*C*)-1,2,4-dithiazole-3-thione
EBDC	ethylene bisdithiocarbamate
EDA	ethylenediamine
EDI	ethylene diisothiocyanate
ETD	ethylene bisthiuram disulfide
ETU	ethylenethiourea
EU	ethyleneurea
ip	intraperitoneal
iv	intravenous
JMPR	Joint Meeting of the FAO Panel of Experts on Pesticide Residues in Food and the Environment and a WHO Expert Group on Pesticide Residues
MIT	methylisothiocyanate
NDDC	sodium diethyldithiocarbamate
NDMA	nitrosodimethylamine
NDMC	sodium dimethyldithiocarbamate
PBI	protein-bound iodine
PTU	propylenethiourea

[a] In some older studies, DIDT is referred to as ethylenethiuram monosulfide (ETM). However, in 1974 the chemical that had been referred to as ETM was shown to be DIDT, and so the latter term has been used throughout this document.

SGPT	serum glutamic-pyruvic transaminase
T_3	triiodothyronine
T_4	thyroxine
TSH	thyroid-stimulating hormone

PART A

DITHIOCARBAMATE PESTICIDES:
A GENERAL INTRODUCTION

CONTENTS

Page

INTRODUCTION

The dithiocarbamates included in this review are those that are mainly used in agriculture and form part of the large group of synthetic organic pesticides that have been developed and produced on a large scale in the last 40 - 50 years. The development of dithiocarbamate derivatives with pesticidal properties occurred during and after the Second World War. However, a few compounds, such as thiram and ziram, were introduced in the 1930s.

The world-wide consumption of dithiocarbamates is between 25 000 and 35 000 metric tonnes per year. Dithiocarbamates are used as fungicides, being effective against a broad spectrum of fungi and plant diseases caused by fungi. In industry, they are used as slimicides in water-cooling systems, in sugar, pulp, and paper manufacturing, and as vulcanization accelerators and antioxidants in rubber. Because of their chelating properties, they are also used as scavengers in waste-water treatment. The herbicidal compounds, which are an integral part of industrialized agriculture, are used mostly in North and Central America, and Europe, with little use reported in Asia, South America, and Africa.

In this introductory document, an attempt has been made to summarize the available data on the dithiocarbamates used as pesticides, in order to indicate their impact on man, animals, plants, and the environment. This overview is not complete, nor is it intended to be. More details on certain aspects are given in the JMPR and IARC (International Agency for Research on Cancer) reports, which have already been published. It also should be recognized that the design of a number of the studies cited, especially the older ones, is inadequate.

1. SUMMARY

1.1 General

Dithiocarbamates are mainly used in agriculture as insecticides, herbicides, and fungicides. Additional uses are as biocides for industrial or other commercial applications, and in household products. Some are used for vector control in public health.

The general formula of dithiocarbamates is characterized by the presence of:

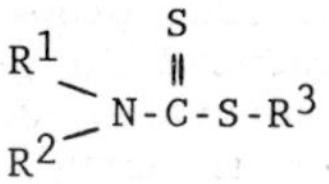

Depending on the types of monoamines used in the synthesis of these compounds, mono- or dialkyldithiocarbamates are formed. Reactions with diamines result in the formation of two terminal dithiocarbamate groups linked by an alkylene (ethylene) bridge. Both alkyl and ethylene dithiocarbamates form salts with metals, and both can be oxidized to the corresponding disulfides.

More than 15 dithiocarbamates are known. However, it is beyond the scope of this publication to give complete information on each compound. The intention is to cover the different aspects of dithiocarbamates, making use of publications and reports available on the compounds that are most used and best known. Data on the carbamates or thiocarbamates are not included, because these compounds have been covered in Environmental Health Criteria 64: Carbamate Pesticides and 76: Thiocarbamate Pesticides (WHO, 1986b; WHO, 1988)

1.2 Properties, Uses, and Analytical Methods

Dithiocarbamates with hydrophylic groups form water-soluble, heavy-metal complexes, while some of the dithiocarbamate metal complexes used as fungicides are insoluble in water but soluble in non-polar solvents. Alkylene bisdithiocarbamates (containing two donor CS_2 groups), which form polymeric chelates, are insoluble in both water and non-polar solvents.

The heavy-metal salts of ethylene bisdithiocarbamic acid may polymerize. Dithiocarbamates may decompose under certain circumstances into a number of compounds, such as sulfur, 5, 6-dihydro-3*H*-imidazol [2,1-C]-1, 2, 4-dithiazole-3-thione, ethylenethiourea (ETU), and ethylenediamine (EDA). ETU is fairly stable, has a high water solubility, and is of particular importance because of its specific toxicity. For this reason, toxicological information on this compound is included in this review.

Physical and chemical data for individual substances are tabulated in the document, and analytical methods for dithiocarbamates are described. Further details for individual dithiocarbamates appear in the WHO Technical Report Series and the IRPTC data profiles.

1.3 Sources, Environmental Transport and Distribution

Most dithiocarbamates were developed during and after World War II. However, a few compounds (ziram and thiram) were introduced around 1931. Dithiocarbamates, with their insecticidal, herbicidal, and fungicidal properties, have a wide range of applications and are produced in great quantities. Because of their high biological activity, dithiocarbamates are also used in medicine, the rubber industry, and in the treatment of chronic alcoholism.

Alkyl dithiocarbamates are stable in an alkaline medium. By splitting off carbon disulfide and hydrogen sulfide, as well as by oxidative degradation, a number of break-down products, such as ETU, are formed in soil and water. The rate of degradation depends on a number of factors, including pH and type of cation. Ethylene bisdithiocarbamates (EBDCs) are generally unstable in the presence of moisture, oxygen, or biological systems, and decompose rapidly in water.

The mobility of EBDCs in soil varies considerably, depending on their individual water solubilities and the type of soil. ETU is water-soluble and mobile. It is taken up by plant roots, is translocated, and metabolized, forming ethyleneurea (EU), other 2-imidazole derivatives, and various unidentified metabolites. In addition, ETU is readily photooxidized to EU in the presence of photosensitizers. Residues of EBDCs and ETU are found in and/or on crops treated with EBDCs. The residue levels change during storage, processing, and cooking due to environmental factors. During these processes, the parent compound may be converted to ETU.

1.4 Environmental Levels and Human Exposure

Information on the environmental impact of dithiocarbamates with respect to persistence and bioaccumulation in the different species and food chains is limited. On the basis of the available information, it is likely that most of these compounds are rapidly degraded in the presence of oxygen, moisture, etc., to form a number of compounds, some of which, e.g., ETU and propylenethiourea (PTU), are toxicologically important.

When certain crops, such as spinach, carrots, and potatoes, are treated with EBDCs, high levels of ETU can be found after cooking. In general, however, the ETU levels are below 0.1 mg/kg product.

Human exposure to EBDCs was calculated for the population of the USA on the basis of estimated consumption of dietary residues of ETU in treated crops. Upper limit (worst case) and lower limit (lowest case) estimates of exposure to ETU were 3.65 µg/kg and 0.24 µg/kg body weight per day, respectively.

An estimate made for the Canadian population on the basis of results of available market-basket surveys would be around 1 µg/kg body weight per day.

1.5 Kinetics and Metabolism

As a general rule, dithiocarbamates can be absorbed by the organism via the skin, mucous membranes, and the respiratory and gastrointestinal tracts. Whereas dithiocarbamates are absorbed rapidly from the gastrointestinal tract, metal-complexed alkylene bisdithiocarbamates are absorbed poorly both from the gastrointestinal tract and through the skin.

Dialkyldithiocarbamates and EBDCs are metabolized via different mechanisms. The metabolism of the former is straightforward, dialkylthiocarbamic acid being formed as a free acid or as *S*-glucuronide conjugate. Other metabolic products include carbon disulfide, formaldehyde, sulfate, and dialkyl amine.

The metabolic decomposition of EBDCs in mammals is complex and results in the formation of carbon disulfide, EDA, a few ethylene bisthiuram disulfides, hydrogen sulfide, ethylene bisthiocyanate, and ETU. The latter is further broken down to moieties that are incorporated into compounds such as oxalic acid, glycine, urea, and lactose. Dithiocarbamates and their metabolic products are found in certain organs, such as the liver, kidneys, and, especially, the thyroid gland, but accumulation of these compounds does not take place because of their rapid metabolism.

After treating plants with dithiocarbamates, a large number of metabolites are found, including ETU, EU, imidazole derivatives, diisothiocyanates, diamines, disulfides, and other metabolites, that are still unknown.

1.6 Effects on Organisms in the Environment

Soil microorganisms are capable of metabolizing dithiocarbamates. From the limited information available, it seems that the breakdown products can affect enzyme activities, respiration, and nitrification at dose levels of the order of 10 mg/kg dry soil or more.

Dithiocarbamates have an LC_{50} of less than 1 mg/litre for invertebrates *(Daphnia)* and between 1 and 4 mg/litre for algae

(*Chlorella*). The acute toxicity of dithiocarbamates for fish is rather high. In general, the acute LC_{50} of dialkyldithiocarbamates for fish is less than 1 mg/litre, and that of EBDCs is in the range 1 - 8 mg/litre water. The sac fry and early fry stages of the rainbow trout have a higher sensitivity than other early life stages, and embryotoxic and teratogenic effects are induced by certain dithiocarbamates. However, bioaccumulation is low (bioconcentration factor < 100). The toxicity of ETU and EU for fish, *Daphnia, Chlorella,* and two bacteria species is very low, of the order of g/litre.

Several dithiocarbamates were shown to intervene with testicular development and function and to cause nerve fibre degeneration in domestic fowl.

Information on the influence of dithiocarbamates on honey bees is lacking.

1.7 Effects on Experimental Animals and *In Vitro* Test Systems

The acute oral and dermal toxicities of the different dithiocarbamates are generally low. Most compounds have a low volatility, and only limited information concerning inhalation toxicity is available. Local irritation of the respiratory tract occurs when dithiocarbamates are inhaled as dust, which can also induce eye and dermal irritation. Some dithiocarbamates are sensitizing agents. ETU also has a low acute oral toxicity.

Many short- and long-term toxicity studies have been carried out on different dithiocarbamates. In rats, some dithiocarbamates tested at high dose levels induced dose-dependent adverse effects on the reproduction and endocrine structures and functions, thus reducing reproductive capacity. Some dithiocarbamates also showed effects on reproduction in birds.

In teratogenicity studies on mice and rats, dithiocarbamates induced an increase in resorption sites and somatic and skeletal malformations (cleft palate, hydrocephaly, and other abnormalities). The dose levels needed to produce these effects were usually higher than 200 mg/kg body weight in rats, and above 100 mg/kg body weight in mice.

In general, the results of mutagenicity studies with dithiocarbamates have been negative.

From the available long-term carcinogenicity studies on mice and rats, there is no clear indication of a carcinogenic effect. Some of the dithiocarbamates have shown a goitrogenic effect at high dose levels.

There is evidence that certain dithiocarbamates may be converted *in vivo* to *N*-nitroso derivatives, which are considered to be both mutagenic and carcinogenic. However, the levels of nitroso compounds that can be expected to result from the dietary intake of dithiocarbamate pesticide residues are negligible compared with those of the nitroso precursors, which occur naturally in food and drinking-water.

In rats, high levels of dithiocarbamates produce an increase in thyroid weight, a reduction in colloid in follicles, hyperplasia, and nodular goitre. These distinct morphological changes are in agreement with an increase in thyroid-stimulating hormone (TSH). Hypophyseal stimulation of the thyroid is the consequence of a decreased blood level of thyroxin, the synthesis of which is inhibited by dithiocarbamates. The thyroid hyperplasia induced by dithiocarbamates is largely reversible on cessation of exposure.

Another intriguing phenomenon is the induction of alcohol intolerance by most of the alkyldithiocarbamates. This phenomenon has been studied in rats and produced in man. It has even led to the use of disulfiram in the treatment of chronic alcoholism.

At dose levels above 50 mg/kg body weight, dithiocarbamates produce neurotoxic effects in rats and rabbits, characterized by ataxia and paralysis of the hind legs, and demyelination and degeneration of peripheral nerves. In birds, paralysis and muscular and peripheral nerve atrophy have also been observed.

Dithiocarbamates have been reported to cause a redistribution of heavy metals, e.g., lead and cadmium, in organs such as the brain. Furthermore, because of their chelating properties, these dithiocarbamates may have an effect on the function of enzymes containing metals, such as zinc and copper.

1.8 Effects on Man

Regular contact with dithiocarbamates can cause functional changes in the nervous and hepatobiliary systems. Skin contact with dithiocarbamates may induce contact dermatitis, and some of these compounds will induce sensitization. Alcohol intolerance can be induced by certain dithiocarbamates, as indicated in section 1.7.

There are indications that the mean incidence of chromosomal aberrations in lymphocytes is increased in workers exposed to certain dithiocarbamates. Epidemiological studies on workers exposed to dithiocarbamates or ETU did not show any increase in the incidence of thyroid tumours. However, only a relatively small number of workers was involved.

2. IDENTITY, PHYSICAL AND CHEMICAL PROPERTIES, ANALYTICAL METHODS

2.1 Identity

Dithiocarbamates are the disulfur analogues of carbamates, and they are characterized by the presence of:

$$>N-\overset{\overset{S}{\|}}{C}-S-$$

Secondary monoamines, e.g., dimethyl or diethyl amines, react with carbon disulfide to give dialkyldithiocarbamates:

$$R_2NH + CS_2 + NaOH \text{ ---- } R_2N-\overset{\overset{S}{\|}}{C}-S-$$

Reaction with monoalkylamines gives the corresponding monoalkyldithiocarbamates. The reaction of carbon disulfide with diamines (for instance, EDA) gives two terminal dithiocarbamate groups linked by an alkylene bridge:

$$(CH_2NH_2)_2 + CS_2 + NaOH \text{ ----> } \begin{array}{l} CH_2-NH-\overset{\overset{S}{\|}}{C}-S- \\ | \\ CH_2-NH-\underset{\underset{S}{\|}}{C}-S- \end{array}$$

Both alkyl and ethylene dithiocarbamates form salts with metals and both can be oxidized to the corresponding disulfides. EBDCs can form polymers, especially in the presence of certain ubiquitous metallic ions (Engst & Schnaak, 1974).

The chemical structures and pesticidal activity of the principal dithiocarbamates are listed in Table 1. CAS registry numbers, chemical names, common names, molecular formulae, relative molecular masses, and selected chemical and physical properties are summarized in Annex I. Further information can be obtained from the JMPR evaluations (Annex III).

2.2 Physical and Chemical Properties

Dithiocarbamates with hydrophylic groups, such as OH^- and COOH, form water-soluble heavy metal complexes. However, dithiocarbamate metal complexes used as fungicides are all insoluble in water, though they are soluble in non-polar solvents. Alkylene bisdithiocarbamates containing two donor

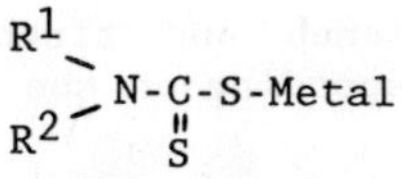

Dithiocarbamate

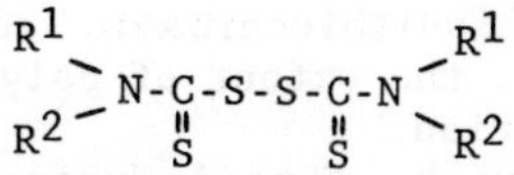

Thiuram disulfide

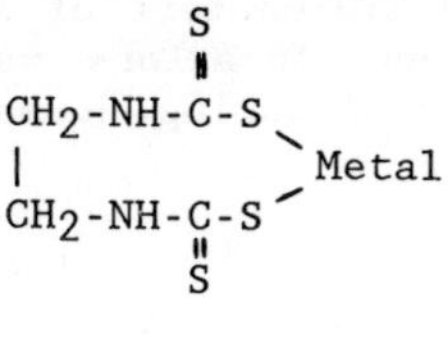

EBDC

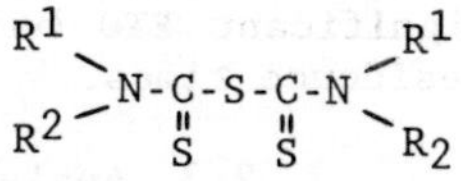

Thiuram monosulfide

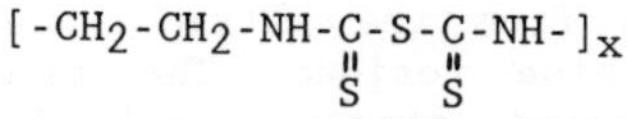

Polymer

Table 1. Relationship of chemical structure and pesticidal activity of dithiocarbamates

Pesticidal activity	Chemical structure	Common or other name
Herbicides	S ‖ dialkyl-N-C-S-alkyl	sulfallate[a]
Fungicides and/ or insecticides	S ‖ >N-C-S-Metal	ferbam, mancozeb, maneb, metam-sodium[b], metiram, nabam, propineb, zineb, ziram

[a] Pre-emergence herbicide.
[b] Soil fungicide, nematocide, and herbicide.

CS_2^- groups, which form polymeric chelates, are insoluble in both water and non-polar solvents.

Dithiocarbamates are unstable in acidic conditions and readily convert to the amine and carbon disulfide (Ludwig & Thorn, 1962; Thorn & Ludwig, 1962). The heavy metal salts of ethylene bisdithiocarbamic acid, i.e., maneb and zineb, may polymerize, the extent of polymerization depending on the method of preparation.

ETU may be formed during the manufacture of dithiocarbamates. Bontoyan & Looker (1973) studied the initial ETU content of various EBDC products and the amount found after storage. Lyman & Lacoste (1974) found that the average ETU content of 76 lots of mancozeb manufactured at six different locations was 0.07%. No significant ETU build-up was observed during normal spray tank residence times.

2.3 Analytical Methods

Residue analysis consists of sampling the environmental material or matrix, extracting the pesticide residue, removing interfering substances from the extract, and identifying and quantifying the pesticide residue. The manner in which the matrix material is sampled, stored, and handled can affect the results: samples should be truly representative, and their handling and storage must not further contaminate or degrade the residue being measured.

The dithiocarbamates, thiuram disulfides included, are conveniently determined on the basis of their decomposition by mineral acids to the amine and carbon disulfide. The amount of either of these hydrolysis products can be determined, the carbon disulfide being commonly measured iodometrically or colorimetrically. This decomposition method is adaptable to micro-determinations for the assay of pesticide residues on crops or to the macro-methods, which are used to determine concentrations of ingredients in pesticide formulations (Clarke et al., 1951).

A polarographic method has been used to estimate residues of maneb and zineb (detection limit, 0.5 mg/kg product) and ethylene bisthiuram monosulfide (detection limit, 0.02 mg/kg product) (Engst & Schnaak, 1969a,b,c, 1970b).

A number of procedures for the quantification of dithiocarbamates are based on high-pressure liquid chromatography. The limits of detection in water solutions for zineb, ziram, and thiram are 0.05, 0.01, and 0.01 mg/kg, respectively (IARC, 1976; Gustafsson & Thompson, 1981; Kirkbright & Mullins, 1984; Tetsumi et al., 1985).

For further details of analytical methods for individual dithiocarbamates, see Conkin & Gleason (1964), Fishbein (1975), Ashworth et al. (1980), and Worthing & Walker (1983).

3. SOURCES OF HUMAN AND ENVIRONMENTAL EXPOSURE

3.1 Natural Occurrence

No data are available.

3.2 Man-Made Sources

The development of mono- and dithiocarbamate derivatives with pesticidal properties occurred during and after World War II. However, a few compounds were introduced earlier, including ziram in 1930 and thiram in 1931.

Dithiocarbamates were developed as practical field fungicides in the United Kingdom in about 1936. The compounds were already being explored as fungicides and insecticides in the USA, where the classic Tisdale and Williams patent was issued in 1934. This covered the use of compounds of the formula $X(Y)NCS_2Z$ (where X is hydrogen or alkyl, Y is hydrogen, alkyl, or aryl, and Z is metallic in nature) and thiuram sulfides as bactericides and fungicides (Thorn & Ludwig, 1962).

3.2.1 Production levels, processes, and uses

Dithiocarbamates have also been used to control various dermatophytes (Kligman & Rosensweig, 1948). For example, tetramethylthiuram disulfide, incorporated in various soaps and lotions, has been used since 1942 for the treatment of scabies and other parasitic diseases of the skin in veterinary and human medicine (Schultheiss, 1957). Dithiocarbamates also have considerable biocidal activity against a number of protozoa.

An interesting development was the discovery of disulfiram as a treatment for chronic alcoholism (Hald & Jacobsen, 1948). Other important applications of dithiocarbamates are in the field of rubber chemistry as antioxidants and accelerators (Thorn & Ludwig, 1962).

Annual production and use figures for a number of dithiocarbamates in various parts of the world are given in IARC (1976); consumption figures are listed in Table 2.

Table 2. Consumption of dithiocarbamate pesticides (in 100 kg)[a]

Area	Dithiocarbamates			
	1974-76	1981	1982	1983
<u>Africa</u>				
Egypt	30			
Zimbabwe		795		
<u>North/Central America</u>				
Canada	10 977			
Mexico	4531	38 350	34 000	33 050
USA		60 000	50 000	
<u>South America</u>				
Argentina		4890	8370	
Uruguay	1454	822	1114	1668
<u>Asia</u>				
Brunei		3	2	2
Cyprus	701	2242	1538	
India	16 193	14 650	17 130	
Israel	4177	3110	3370	3580
Jordan		27 500	28 748	
Korea Republic	5027	18 380	18 233	
Kuwait	6			
Oman		115	62	120
Pakistan	24	370	881	
Turkey	5906	8901	9346	
<u>Europe</u>				
Austria	2751	2334	2322	2207
Czechoslovakia	6927	8678	6501	
Denmark	2187		11 485	13 747
Finland	504			
Greece	12 763			
Hungary	37 347	29 476	31 932	43 415
Italy	145 697	121 808	97 238	
Malta		350		
Norway	438	383	372	285
Poland	4007	11 386	14 102	12 517
Portugal	8114	8358	7592	
Sweden	3283	3800	4380	

[a] From: FAO (1985).

4. ENVIRONMENTAL TRANSPORT, DISTRIBUTION, AND TRANSFORMATION

Dithiocarbamates, like all pesticides, can reach the soil through many routes, ranging from direct application to drift from foliage treatment. Generally, these compounds are not persistent and undergo different types of degradation.

4.1 Transport and Distribution Between Media

In alkaline medium, alkyl dithiocarbamates are stable, but EBDCs are not. EBDCs are also unstable in the presence of moisture and oxygen as well as in biological systems. By splitting off carbon disulfide and hydrogen sulfide, as well as by oxidative degradation, a great number of secondary products are formed, amongst them, ETU (Aldridge & Magos, 1978).

The rates of alkyl dithiocarbamates decomposition depends on pH (Turner & Corden, 1963) and the cation present. The rate of decomposition, and the production of carbon disulfide is decreased by cations in the following order: $Na^{+} > Zn^{2+} > Fe^{3+} > Cu^{2+}$.

The release of carbon disulfide from EBDCs is influenced by the chemical nature of the hydrolysing medium. It is low in acetic acid and nearly 100% in sulfuric acid (Aldridge & Magos, 1978); it also depends on the temperature (Clarke et al., 1951).

Decomposition to hydrogen sulfide seems to depend on the presence of an N-H group. Monoalkyldithiocarbamates, such as EBDCs, are not stable in alkaline medium and, in acidic medium, decompose either to carbon disulfide or hydrogen sulfide (Joris et al., 1970). The rate of decomposition to carbon disulfide is two orders of magnitude lower in monoalkyldithiocarbamates compared with that in the corresponding dialkyldithiocarbamate (Zuman & Zahradnik, 1957). In the case of metiram sodium at pH 9.5, methylisothiocyanate (MIT) and sulfur are formed, whereas in acid solution, the compound is decomposed into carbon disulfide, hydrogen sulfide, *N,N'*-dimethylthiuram disulfide, methylamine, and MIT (Turner & Corden, 1963).

4.1.1 Water

EBDCs decompose rapidly in water, mancozeb having a half-life of less than 1 day in sterile water (pH range, 5 - 9). The nature and abundance of the degradation products are pH-dependent, and include ETU and EU (Lyman & Lacoste, 1974, 1975). Photolytic degradation is a major pathway for ETU in water (Cruickshank & Jarrow, 1973; Ross & Crosby, 1973), and is enhanced by the presence of photosensitizers such as chlorophyll (Ross & Crosby, 1973).

The half-life of thiram in water was 46.7 days at pH 7 and 9.4 h in an acid medium (pH 3.5). About 5.2% of a sample of thiram was still present in water of pH 7 after 200 days.

4.1.2 Soil

The mobility of EBDCs in soil varies considerably, depending on water solubility and soil type. They are generally more mobile in wet and in sandy soils than in dry soil or soil rich in organic matter (peat or muck). Thin-layer chromatography studies have shown that nabam is more mobile than maneb, which in turn is more mobile than zineb, zineb being almost immobile (Helling et al., 1974).

The leaching of radioactive ^{14}C-mancozeb and its degradation products was studied in five different soils, the organic content of which ranged from 0.4% to 15%, while the pH ranged from 4.7 to 7.4. An aqueous slurry of ^{14}C-mancozeb (15.6 mg) was mixed with a soil sample and applied to the top of a column of soil. Water (2.5 cm) was added to the top of the column once a week for 9 weeks. The water was collected and its radioactivity measured and, after 9 weeks, the columns were cut into 2.5 cm sections. The results showed that no radioactivity leached through four of the five columns (only 2 - 5% of the activity leached through the Cecil clay column; the reason for this is not known). Losses of radioactivity by volatilization or by metabolism to carbon dioxide were significant in all soils (Lyman & Lacoste, 1974).

4.2 Biotransformation

4.2.1 Microbial degradation

Sterilized and unsterilized samples of sewage, fresh water, sea-water, and agricultural soil were incubated with 50 or 100 mg thiram per litre or kg. Thiram disappeared from sewage and fresh water within 12 days, and from soil after 40 days. After 8 months, 20% of the thiram was still present in sea-water. Disappearance was faster in unsterilized than in sterilized soil, indicating that microorganisms seem to be involved (Odeyemi, 1980).

The results of a study on one soil (Hagerstown silt loam) used in the leaching study mentioned in section 4.1.2, showed that mancozeb is readily degraded by soil microorganisms, releasing ethylene C atoms as carbon dioxide. No carbon dioxide was released from sterile soil, but mancozeb was rapidly degraded to carbon dioxide in non-sterile soil. The half-life in soil at a concentration of 20 mg mancozeb/kg was 50 days; at 10 mg/kg, the half-life was 90 days (Lyman & Lacoste, 1974).

4.2.2 Photodegradation

Ziram is stable to ultraviolet radiation (UVR). It is slowly photo-hydrolysed in water and is stable in media containing quantities of organic acids. When precipitated to the bottom of bodies of water, it remains toxic for a month (IRPTC, 1982).

5. ENVIRONMENTAL LEVELS AND HUMAN EXPOSURE

The only exposure of the general population to dithiocarbamates and their breakdown products results from occasional residues in the diet. However, dithiocarbamates degrade rapidly after application to crops, the rate being influenced by oxygen, humidity, temperature, organic sensitizers, and pH. A number of degradation products have been identified, including ETU, ethylene thiuram disulfide (ETD), and DIDT[a].

5.1 Food

Studies in Canada and the USA have shown that, when vegetables, such as spinach, carrots, and potatoes, are treated with EBDCs after harvest, a significant percentage of the EBDCs is converted to ETU during subsequent cooking (Blazquez, 1973; Newsome & Laver, 1973; Watts et al., 1974) (ETU section 5.1).

The results of a study by Phillips et al. (1977) to examine the effects of food processing on EBDC residues confirmed and extended the results described above. Washing the raw agricultural products prior to processing removed 33 - 87% of the EBDC residues and the majority of the ETU residues. The results for raw and processed commodities are summarized in Table 3.

Human exposure to EBDCs was calculated for the population of the USA on the basis of estimated consumption of dietary residues of ETU in treated crops. Upper limit (worst case) and lower limit (lowest case) estimates of exposure to ETU were 3.65 μg/kg and 0.24 μg/kg body weight per day, respectively (US EPA, 1982b).

EBDC residues would be expected to be lower in root crops, such as carrots and potatoes, as they are not systemic and tend to remain on the external portions of the plant. However, in leafy crops, such as spinach and lettuce, EBDC residues are generally higher. Culling, such as discarding the discoloured leaves of lettuce and the rinds of melons, could presumably reduce the residue level. Washing reduced the majority of EBDC residues by at least 50%.

---

[a] In many publications, it has been stated that ethylenethiuram monosulfide (ETM) was identified in metabolic studies; however, it is now clear that this metabolite is 5, 6-dihydro-3*H*-imidazo [2, 1-C]-1,2,4-dithiazole-3-thione (DIDT) (Pluygers et al., 1971; Benson et al., 1972; Alvarez et al., 1973).

Table 3. Summary of EBDC/ETU residues (mg/kg product) before and after processing

	Eastern USA		Western USA	
	EBDC	ETU	EBDC	ETU
Tomatoes				
Unwashed	0.3	-	2.1	0.01
Washed	0.2	-	0.6	0.01
Canned	-	0.03	0.5	0.11
Carrots				
Unwashed	0.6	-	0.1	0.01
Washed	0.3	-	0.1	0.01
Diced	0.1	-	0.1	-
Frozen	-	-	-	-
Canned	-	0.03	0.1	-
Spinach				
Unwashed	2.4	-	61.9	0.34
Washed	1.5	-	9.7	0.02
Frozen	0.1	0.04	0.6	0.50
Canned	-	0.18	0.1	0.71

Note: Mancozeb was applied at the rate of 0.7 ai/0.5 ha in all cases. Spray schedules were as follows: spinach, 1 treatment with 10-day pre-harvest interval; carrot, 6 treatments at 7- to 10-day intervals (7-day pre-harvest interval); tomato (eastern), 4 treatments at 7- to 10-day intervals (16-day pre-harvest interval); tomato (western), 3 treatments at 7-day intervals (5-day pre-harvest interval). From: IUPAC (1977).

5.2 Monitoring and Market Basket Studies

In a market-basket study, over 500 samples of 34 foods were analysed, together with 26 samples of drinking-water. The water samples and 338 food samples did not contain any residues. Doubtful positive values at approximately the limit of detection were found in 110 food samples, and 53 samples were positive. Only 21 of all the samples contained ETU residues.

Tomato products (203 samples) were analysed in a separate market-basket study, and 19% contained dithiocarbamates in the range of 0.2 - 0.5 mg/kg product (Gowers & Gordon, 1980).

A more realistic review of the actual exposure of the general population was obtained by a "table-top" study in which 100 whole meals (60 from homes and 40 from restaurants) were analysed for dithiocarbamates and ETU. In the 87 meals analysed for dithiocarbamates, 11 contained residues of apparent dithiocarbamates averaging 0.3 mg/kg, or 0.04 mg/kg if averaged over the 87 meals. In a second study of 100 meals, 4 meals contained apparent dithiocarbamates in the range of 0.2 - 0.4 mg/kg or 0.02 mg/kg as an average of the 100 meals. From these studies, an overall average would be 0.03 mg dithiocarbamates/kg meal. ETU residues were not found in either study (Gowers & Gordon, 1980).

6. KINETICS AND METABOLISM

Dithiocarbamates penetrate the organism mainly via the respiratory tract (aerosol, dust), skin and mucous membranes (occupational exposure), and the digestive tract.

6.1 Absorption, Distribution, and Excretion

Thirty minutes after intragastric administration of 500 mg ziram/kg body weight to rats, the compound was detected in the blood, the liver, and the kidneys, the highest concentration being in the liver (26.2 mg/kg tissue). After 16 h, the concentration of ziram in the blood and liver (about 5.5 mg/kg tissue) decreased considerably, while the concentration in the intestines and the kidneys increased (in the kidneys, to 3 mg/kg tissue). At the end of the first day, the ziram concentration in the intestines reached a maximum and then dropped abruptly, 57% of unchanged ziram being detected in the faeces; the compound was also detected in the spleen and the adrenal glands. Maximum concentrations in the organs (6.8 mg/kg and 2.4 mg/kg, respectively) were attained the following day. Ziram was no longer present in the adrenal glands after 3 days, and in the spleen after 6 days. The circulation of ziram in the blood continued for 2 days (Vekshtein & Khitsenko, 1971).

After the oral administration of a dose of 2 mg ^{35}S-ziram per animal to white rats (100 - 120 g), the brain and thyroid contained high levels of radioactivity during the first 2 days. During the 12 h following administration, higher amounts of ziram (or its metabolites) were found in the ovaries than in the uterus or the placenta. Ziram passed the placental barrier and accumulated in the organs and tissues of the fetus (skin, liver, heart) at levels several times higher than those in the placenta and the uterus wall. The level of radioactivity in the fetal liver exceeded the maximum level in the liver of mature animals; at 12 h, it was more than 5 times higher (Chernov & Khistenko, 1973). Twenty-four hours after administering ^{35}S-ziram to female rats, Izmirova & Marinov (1972) found radioactivity in the thyroid, blood, kidneys, spleen, ovaries, and liver.

When ^{14}C-labelled maneb was orally administered to rats at a dose of 360 mg/kg body weight by stomach tube, approximately 55% of the radioactivity was eliminated in the faeces and urine within 3 days. Almost no unmetabolized maneb was found. The amounts of radioactivity in organs after day 1 and day 5 were 1.2% and 0.18%, respectively. The highest levels after 1 day were found in blood (0.23%), liver (0.78%), and kidneys (0.18%). Less was found in the thyroid (0.07%) (Seidler et al., 1970).

Similar results were obtained with rats administered ^{35}S-ferbam or ^{14}C-ferbam. Approximately 50% was absorbed from the gastrointestinal tract in the first 24 h. Rats receiving ^{35}S-ferbam showed 18%, 23%, and 1% in expired air, urine, and bile, respectively, whereas with ^{14}C-ferbam, the figures were < 0.1%, 43%, and 1.4%, respectively. Other tissues contained only small amounts of labelled material. In addition, ^{14}C was excreted in the milk of lactating rats (Hodgson et al., 1974).

Blackwell-Smith et al. (1953) found that approximately 70 - 75% of ingested zineb passed through the gastrointestinal tract of rats and appeared in the faeces within 24 - 72 h.

Rats dosed via a stomach tube with 20 mg ^{14}C-mancozeb per day for 7 days (equivalent to approximately 100 mg/kg body weight) were killed one day after the last dose and the radioactivity in excreta and organs was measured. In the faeces, urine, organs and tissues, and carcass, 71%, 16%, 0.31%, and 0.96% of the total radioactivity was detected, respectively. Specifically, the liver contained 0.19%, the kidneys, 0.076%, the thyroid gland, 0.003%, and all other organs, less than 0.01%. Most of the labelled material in the faeces was mancozeb, indicating that mancozeb was poorly absorbed from the gastrointestinal tract (Lyman, 1971).

6.2 Metabolic Transformation

Dialkyldithiocarbamates, such as thiram and disulfiram, and EBDCs, such as nabam, maneb, and zineb, are metabolized via different mechanisms.

6.2.1 Mammals

In general, the metabolism of dialkyldithiocarbamates (e.g., disulfiram) in mammals (including man) is straightforward, diethylthiocarbamic acid being formed as the principal metabolite. This is found either as the free acid or as the *S*-glucuronide conjugate (Fig. 1) (Kaslander, 1963; Strömme, 1965; Dekhuyzen et al., 1971; Aldridge & Magos, 1978) in the urine, faeces, or tissues of animals. Other metabolic products include carbon disulfide (Prickett & Johnston, 1953), methyldiethyldithiocarbamate (Gessner & Jakubowski, 1972), and sulfate (Strömme, 1965; Strömme & Eldjarn, 1966), but free disulfiram was not detected.

One of the most important enzymatic processes in the metabolism of dialkyldithiocarbamates is glucuronidation, which takes place in the liver (Strömme, 1965). Glucuronic acid conjugation might be overloaded after the administration of diethyldithiocarbamate but not after the administration of disulfiram, which is taken up by the liver at a much slower rate. Methylation of diethyldithiocarbamates by *S*-adenosyl methionine transmethylase in the kidneys and liver can occur subsequently and leads to sulfate excretion (Gessner &

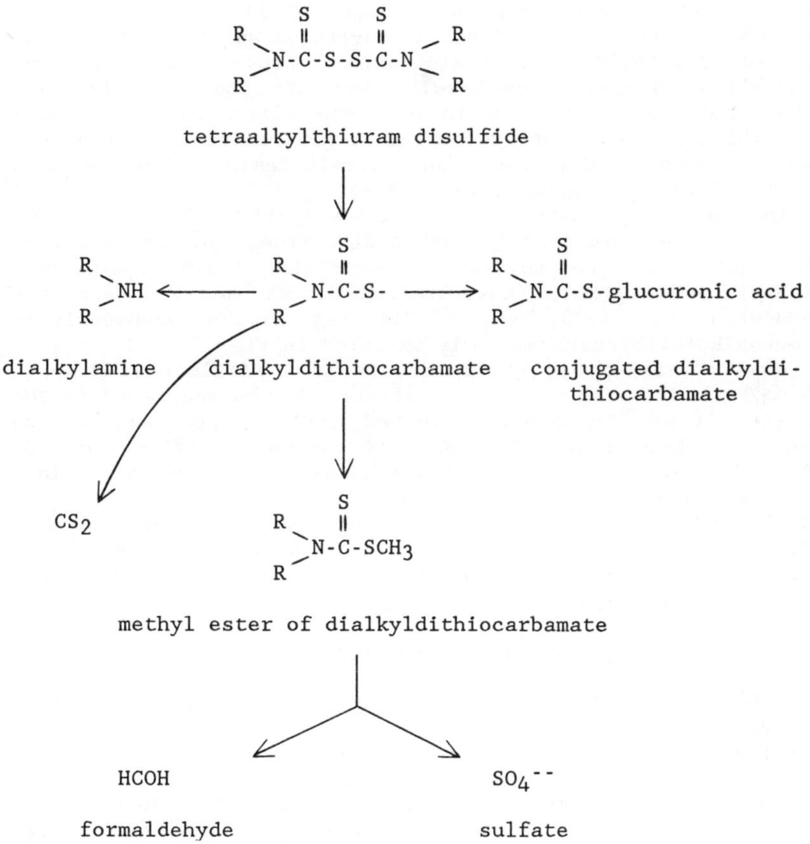

Fig. 1. Metabolic pathways for the decomposition of dialkyldithiocarbamates. From: Aldridge & Magos (1978).

Jakubowski, 1972). In the case of ^{35}S-disulfiram, more than 50% of the ^{35}S was recovered as sulfate in the urine, partly in the free form and partly esterified (Eldjarn, 1950; Strömme, 1965). A different enzymatic process is involved in the desulfuration of the carbon disulfide formed from dithiocarbamates. After administration of ^{14}C-carbon disulfide, some label was exhaled as ^{14}C-carbon dioxide. This break-up of the carbon disulfide molecule is catalysed by microsomal mixed-function oxidase (De Matteis & Seawright, 1973; Dalvi et al., 1974).

Thiram and the dimethylamine salt of dimethyldithiocarbamic acid were the major metabolites in the urine, whereas carbon disulfide and dimethylamine were detected in the expired air. The body tissues contained tetramethylthiourea, the methylamine salt of dimethyldithiocarbamic acid, carbon disulfide, and methylamine. Overall, the results indicate that, in the rat, ferbam and ziram are transformed into dimethyldithiocarbamic acid, which is subsequently coupled to give thiram, or is broken down to carbon disulfide and dimethylamine (Vekshtein & Khitsenko, 1971; Hodgson et al., 1974).

The *in vivo* metabolic decomposition of EBDCs is complex and results in the formation of carbon disulfide, hydrogen sulfide, EDA, ethylene bisthiuram disulfide, DIDT, ethylene diisothiocyanate (EDI) (unstable), ETU, EU, and 2-imidazoline (Seidler et al., 1970; Lyman, 1971) (Fig. 2). The decomposition of monoalkyldithiocarbamates is detailed in Fig. 3.

When ^{14}C-maneb was given to rats in a single oral dose of 390 mg/kg body weight, only 55% of the ^{14}C was recovered in the excreta. It was therefore suggested that a large part of the dose might have been metabolized to carbon disulfide and ^{14}C-EDA, followed by oxidation of the latter to carbon disulfide. The concentration of the radioactivity was highest after 24 h, and EDA and ETU were identified in the excreta (Seidler et al., 1970). ETU and DIDT were the major metabolites found in the urine of rats treated with zineb, and carbon disulfide was detected in the expired air.

6.3 Metabolism in Plants

ETU is one of several metabolites found when EBDCs are applied to plants. In plants, nabam, maneb, and zineb are transformed to ETU, DIDT, EU, 2-imidazoline, a diisothiocyanate (EDI), and other metabolites (Fig. 2).

Nash & Beall (1980) have studied the fate of maneb and zineb in microagroecosystem chambers (enclosed glass chambers), under the following conditions: pH, 6.7; organic matter content, 5.2%; soil type, Galestown sandy loam; soil water content, 15.6%. The fungicides were applied twice to tomato plants at 2 kg/ha, and the residual fungicides (measured as EDA and ETU) were monitored on the fruit, leaves, and in the soil, water, and air for 100 days after treatment. ETU was detected at < 20 $\mu g/kg$ on whole fruit after 3 days, but had completely disappeared after 3 weeks. Maneb and zineb were present on whole fruit at < 1 mg/kg and were still present in measurable amounts (as EDA) after 10 weeks. Both had half-concentration times ($C_{1/2}$) of 14 days on leaves. Half-concentration times for ETU, maneb, and zineb in soil were < 3, 36, and 23 days, respectively, and that for ETU in air was 9 days.

Besides ETU, other degradation products of EBDCs include ETD and DIDT. The main volatile components of zineb's decomposition are carbon disulfide, carbonyl sulfide, and EDA. Almost all

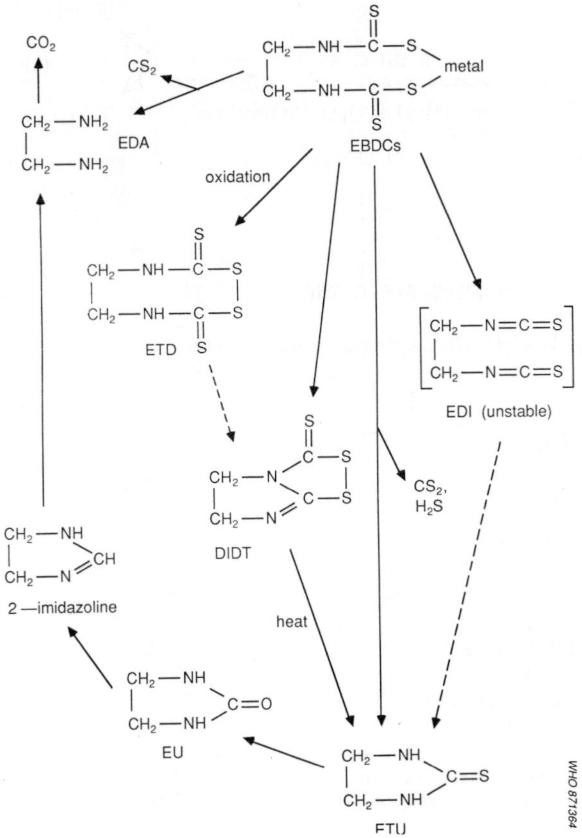

Fig. 2. Metabolic pathways for the decomposition of ethylene bisdithiocarbamates and reactions leading to ETU. Adapted from: Engst & Schnaak (1967, 1970a), Freudenthal et al. (1977), and Aldridge & Magos (1978).

Note: Ion mechanisms leading to ETU formation are not completely understood; however, a number of hypotheses have been advanced. According to Marshall (1977), intermediary products of the thermal bisdithiocarbamate degradation to ETU are β-amino ethylene dithiocarbamate and DIDT, but not ethylene diisothiocyanate (EDI). EDI was, however, postulated and detected several times as a secondary reaction product of the ethylene bisdithiocarbamate degradation at normal temperatures (Engst & Schnaak, 1970a). ----> indicates a postulated conversion.

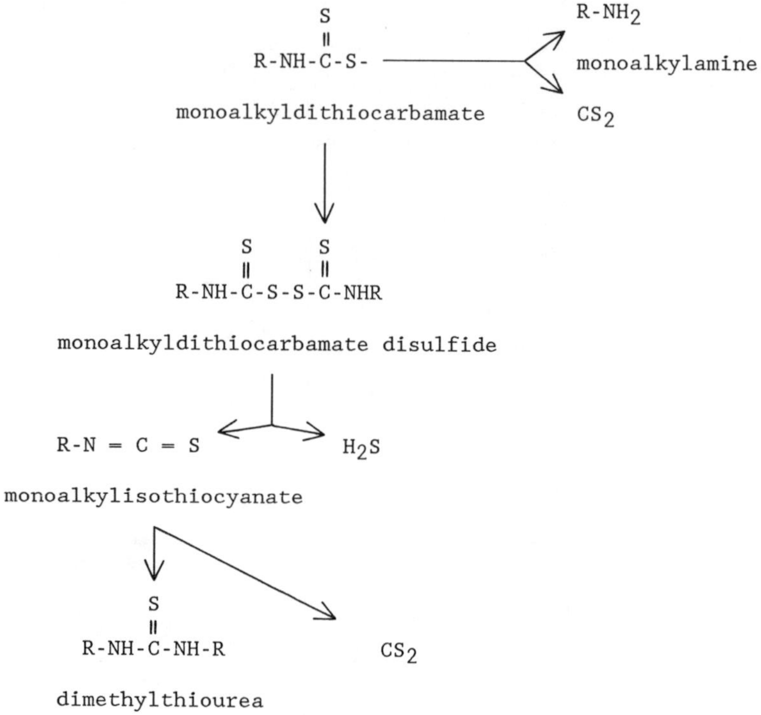

Fig. 3. Metabolic pathways for the decomposition of monoalkyldithiocarbamates (conjugation with glucuronic acid is possible but not shown). From: Aldridge & Magos (1978).

zinc is converted into zinc sulfide and zinc oxide (Melnikov & Trunov, 1966).

6.4 Decomposition in Water and Soil

Thiram and dimethyldithiocarbamic acid give rise in soil to methyl isothiocyanate and sulfur, and, under acidic conditions, to carbon disulfide, hydrogen sulfide, methylamine, methylisocyanate, and the bisdisulfide of methyldithiocarbamic acid. Two of the products, carbon disulfide and dimethylamine, evaporated from the soil (Raghu et al., 1975). Dimethyldithiocarbamic acid also binds with heavy metals in soil to form complexes.

The various metal derivatives of ethylene bisdithiocarbamic acid appear to be converted in the soil to DIDT, ETU, carbon disulfide, hydrogen sulfide, and carbonyl sulfide (Moje et al., 1964; Kaars Sijpesteijn & Vonk, 1970). The conversion by soil bacteria and fungi of DIDT into ETU has been demonstrated (Vonk & Kaars Sijpesteijn, 1976). Even though ETU is slowly converted into EU in soil, pure cultures of soil bacteria and fungi were unable to effect this transformation (Kaars Sijpesteijn & Vonk, 1970).

6.5 Metabolism in Microorganisms

Microorganisms readily form ETU from DIDT, a spontaneous decomposition product of EBDCs. This conversion also takes place after addition of reducing compounds such as cysteine, glutathione, or ascorbic acid. It consists of the reduction of the S-S bond of DIDT, with the subsequent release of carbon disulfide to form ETU. It was shown by Vonk & Kaars Sijpesteijn (1976) that DIDT was reduced by NADH in the presence of enzyme extracts from *Pseudomonas fluorescens*, *Escherichia coli*, *Saccharomyces cerevisiae*, or *Aspergillus niger*.

7. EFFECTS ON ORGANISMS IN THE ENVIRONMENT

7.1 Microorganisms

There is some evidence that dithiocarbamates, at concentrations 10 times that of normal field application, may reduce microbial biomass and increase the bacterial:fungal ratio.

7.2 Aquatic Organisms

7.2.1 Acute toxicity

Toxicity studies using dithiocarbamates are hindered by the fact that they are chemically and biologically degradable and may also be contaminated with degradation products. Their stability in water depends on the pH and on the presence of metal ions with which they form complexes. The soluble dithiocarbamates dissociate in water, whereas the polymers are only slightly soluble in water. As the breakdown products will also influence toxicity, toxicity testing of dithiocarbamates is complex.

According to US EPA (1977, 1982a), the use of pesticide products containing maneb against cranberry fruit rot at application rates of up to 6.7 kg ai/ha would result in a concentration of 4.4 mg/litre in a 15-cm layer of water. McCann & Pitcher (1973) reported a 96-h LC_{50} of 1 mg/litre for bluegills, while Worthing & Walker (1983) reported a 48-h LC_{50} for carp of 1.8 mg/litre. Zineb used against cranberry fruit rot at an application rate of up to 5.4 kg ai/ha could result in a concentration of 3.52 mg/litre in a 15-cm layer of water. A 26-h LC_{50} of 0.2 mg/litre has been reported for *Daphnia magna*.

Van Leeuwen (1986) carried out an extensive study with 18 dithiocarbamates and three metabolites of these compounds, including ETU, in fish *(Poecilia reticulata)*, crustacea *(Daphnia magna)*, algae *(Chlorella pyrenoidosa, Phytobacterium phosphoreum)*, and two nitrifying bacteria *(Nitrosomonas* and *Nitrobacter)*. The results are summarized in Tables 4 and 5.

Worthing & Walker (1983) gave the following LC_{50} values: propineb: rainbow trout, 1.9 mg/litre; golden orfe, 133 mg/litre; thiram: carp, 4 mg/litre; rainbow trout, 0.13 mg/litre; bluegill, 0.23 mg/litre; metiram: harlequin fish, 17 mg/litre.

The susceptibility to maneb of the early life stages of rainbow trout has been studied using fertilized eggs (before and after water hardening), early eye point eggs, late eye point eggs, sac fry, and early fry. The sac fry and early fry stages appeared to be the most sensitive. The 96-h LC_{50}s for the different stages were: for 0-h egg, 6 mg/litre; for 24-h egg,

Table 4. Acute toxicity of dithiocarbamates and breakdown products for fish[a]

Organism	Compound	96-h LC_{50} (95% confidence limit) (mg/litre)
Poecilia reticulata[b]	nabam	5.8 (4 - 8.5)
	maneb	3.7 (3.2 - 5.6)
	zineb	7.2 (5 - 10.3)
	mancozeb	2.6 (2.1 - 3.3)
	metiram	6.4 (4 - 10.4)
	Na-DMDC	2.6 (2.1 - 3.2)
	ziram	0.75 (0.56 - 1)
	ferbam	0.09 (0.06 - 0.18)
	thiram	0.27 (0.22 - 0.33)
	Na-DEDC	6.9 (5.5 - 8.5)
	Zn-DEDC	0.49 (0.40 - 0.61)
	disulfiram	0.32 (0.24 - 0.43)
	ETU	7500 (5600 - 10 000)
	EU	13 000 (10 000 - 18 000)
Rainbow trout[c] *(Salmo gairdneri)*	thiram	0.26 (0.24 - 0.32)

[a] From: Van Leeuwen (1986).
[b] Studies according to OECD guidelines 203.
[c] 24-h LC_{50}; water temperature 15 ± 1 °C; weight of fish, 34 ± 4.7 g.

Table 5. Short-term toxicity studies with dithiocarbamates and breakdown products[a]

Compound	*Daphnia magna* 48-h LC_{50} (mg/litre)	*Chlorella pyrenoidosa* 96-h EC_{50} (mg/litre)	*Photobacterium phosphoreum* 15-min EC_{50} (mg/litre)	*Nitrosomonas Nitrobacter* 3-h MIC (mg/litre)
Nabam	0.44	2.4	102	32
Maneb	1	3.2	1.2	56
Zineb	0.97	1.8	6.2	18
Mancozeb	1.3	1.1	0.08	32
Metiram	2.2	1.8	0.37	32
Na-DMDC	0.67	0.8	0.51	26
Ziram	0.14	1.2	0.15	100
Ferbam	0.09	2.4	0.20	10
Thiram	0.21	1	0.10	18
Na-DECD	0.91	1.4	1.22	43
Zn-DEDC	0.24	1.1	1.70	> 320
Disulfiram	0.12	1.8	1.21	> 320
ETU	26.4	6600	2100	1
EU	5600	16 000	3300	1000

[a] From: Van Leeuwen (1986).

5.6 mg/litre; for early eyed egg (14 days), 1.8 mg/litre; for late eyed egg (28 days), 1.3 mg/litre; for sac fry (42 days), 0.32 mg/litre; and, for early fry (77 days), 0.34 mg/litre (Van Leeuwen, 1986).

7.2.2 Short- and long-term toxicity and reproduction studies

7.2.2.1 Fish

In sublethal toxicity studies carried out on *Salmo gairdneri*, groups of 10 fish were exposed to thiram (0.18 mg/litre) for 24 h. Blood parameters (decreased haemoglobin and leukopenia, decreased glucose levels, and increased glucose-6-phosphate dehydrogenase activity) and liver parameters (increased lipid content, increased lactate dehydrogenase) were changed, and it was concluded by the author that thiram is a cytotoxic chemical (Van Leeuwen, 1986). In a further study, the 60-day toxicity for early life stages was tested on *S. gairdneri* using a number of dialkyldithiocarbamates, EBDCs, and breakdown products. The LC_{50}s ranged from approximately 1 to 9 µg/litre for dialkyldithiocarbamates and 211 - 2100 µg/litre for EBDCs, but ETU and EU were not toxic, even at levels exceeding 1000 mg/litre.

Embryotoxic and teratogenic effects were also observed for all the compounds studied, and there was an overlap between the responses for skeletal malformations and lethality over a wide concentration range. The teratogenic effects in rainbow trout proved to be in agreement with those observed in mammals. Exposure of rainbow trout during embryo-larval development revealed that malformations induced by dithiocarbamates were almost exclusively confined to the notochord, which increased considerably in both length and diameter. As a result, the notochord became twisted and distorted. Ectopic osteogenesis was observed in almost every affected notochord. Other effects, such as the disruption of the integrity of myomeres and organ dislocations, were closely related to the notochordal anomalies. Also, compression and fusion of vertebrae and "waviness" of various skeletal elements were found. Concentration-related changes in the liver were observed in short-term exposure of juvenile rainbow trout, while at high levels proliferation of bile duct epithelial cells and necrosis of hepatocytes were seen. Ziram and thiram induced brain haemorrhages as well as intraspinal extravasates of blood cells (Van Leeuwen, 1986).

7.2.2.2 Invertebrates

Short-term toxicity studies using the compounds listed in Table 5 were carried out to investigate the effects of prolonged exposure (21-day) on survival, fecundity, and growth of *Daphnia magna*. Growth and reproduction were not specifically inhibited, since effects on these characteristics were generally detected at levels comparable with the 21-day LC_{50}s. For dialkyldithiocarbamates, the 21-day LC_{50}s ranged from approximately 10 to 30 µg/litre, for EBDCs, from 80 to 110 µg/litre, and, for ETU and EU, the levels were 18 and 3200 mg/litre, respectively (Van Leeuwen, 1986).

7.2.3 Bioaccumulation

Van Leeuwen (1986) determined the log *n*-octanol/water partition coefficient for a few dithiocarbamates and breakdown products. The results are summarized in Table 6. In general, the higher the log *n*-octanol/water partition coefficient, the greater the tendency to bioaccumulate.

Table 6. log *n*-octanol/water partition coefficients for some dithiocarbamates

Compound	log *n*-octanol/water partition coefficient
Disulfiram	4
Thiram	1.82
ETU	0.67
EU	0.96

In short-term studies on rainbow trout of uptake, distribution, and retention of ^{14}C-labelled zineb and ziram, both compounds were found to be rapidly distributed throughout the tissues. Whole-body accumulation was low, with bioconcentration factors of less than 100. Relatively high radioactivity levels were found in the liver, gall bladder, and intestinal contents, suggesting the prominent role of hepatic biotransformation and biliary excretion. With ziram, the eyes and skin also appeared to be distribution sites. Whole-body elimination was rapid, with about 75% of radioactivity being eliminated within the first 4 days. With ziram, 45% of the initial total ^{14}C content in the body was still present at the end of the 16-day depuration period. Differences in the extent of elimination were most noteworthy for the eyes, skin, and kidneys. Whole-body autoradiography showed the radioactivity in the digestive tract, liver, bile, gills, thyroid follicles, melanophores of the skin, choroid epithelium complex of the eyes, and in other melanin-containing tissues, such as the kidneys (Van Leeuwen, 1986).

8. EFFECTS ON EXPERIMENTAL ANIMALS AND *IN VITRO* TEST SYSTEMS

8.1 Single Exposures

In general, the toxicity of dithiocarbamates for mammals is relatively low. Some commonly used dithiocarbamates included in the WHO recommended classification of pesticides by hazard (WHO, 1986a), which is based primarily on the acute oral and dermal toxicity of the technical material for the rat, are given in Annex III.

Acute oral and dermal toxicity data for a number of animal species of various dithiocarbamates are given in Table 7. From this Table, it is clear that nabam and metam-sodium are the most toxic dithiocarbamates, other compounds having only low toxicity. As for many compounds, the toxicity is often influenced by the method of application, e.g., solvents used, age and sex of animal, type of diet, etc. Thus, in rats fed an isocaloric diet containing 3.5% protein, the LD_{50} of nabam was 210 mg/kg body weight, compared with 565 mg/kg in rats fed the same diet plus 26% protein (Periquet & Derache, 1976).

Ivanova-Chemishanska (1969) found that rats treated with zineb, maneb, or mancozeb showed dose-dependent signs of depression, adynamia, decreased tonus, disturbances in coordination, paresis, and paralysis of extremities combined with general weakness, lack of appetite, and prostration.

Yin-Tak Woo (1983) has reviewed the structure-activity relationships of different types of dithiocarbamates.

8.2 Short- and Long-Term Exposures

8.2.1 Oral exposure

8.2.1.1 Rat

In 1-month feeding tests, no growth retardation was noted in rats fed diets containing 100 mg ferbam/kg diet, but decreased growth occurred with 500 mg/kg diet and increased mortality with 5000 mg/kg diet (Hodge et al., 1952).

Groups of 40 weanling rats (20 females and 20 males) were given diets containing 500, 1000, 2500, 5000, or 10 000 mg zineb/kg diet for up to 30 days. Thyroid enlargement was seen at all dose levels, but unequivocal histopathological changes were observed only at 10 000 mg/kg diet (Blackwell-Smith et al., 1953; Kampmeier & Haag, 1954).

Ferbam administered daily at 23, 66, or 109 mg/kg body weight to male rats for 13 weeks caused death and weight loss at the highest dose, but did not have any effect on reproduction. Daily feeding (equivalent to 15 or 51 mg/kg body weight) to

Table 7. Acute toxicity (LD_{50}) of dithiocarbamates for experimental animals

Compound	Animal	Dose (mg/kg body weight) oral	dermal	Reference
Ferbam	mouse	1000		FAO/WHO (1965b)
	rat	> 4000		Hodge et al. (1956)
	guinea-pig	450 - 2000		
	rabbit	2000 - 3000		
Metham-sodium (vapam)	mouse	285		Worthing & Walker (1983)
	rat	1700 - 1800		Worthing & Walker (1983)
	rabbit		1300	Worthing & Walker (1983)
Ziram	rat	1400		Hodge et al. (1952)
	guinea-pig	100 - 150		Hodge et al. (1952)
	rabbit	100 - 1020		Hodge et al. (1952)
Thiram	mouse	1500 - 2000		Worthing & Walker (1983)
	rat	865 - 1300		Van Esch (1956)
	rat	780 - 865		Worthing & Walker (1983)
	rat		> 2000	Ben Dyke et al. (1970)
	rabbit	210		Lehman (1951)
	cat	230		
Disulfiram	rat	> 4000		Van Esch (1956)
Zineb	rat	> 5200		Blackwell-Smith (1953)
	rat	9000		Ivanova-Chemishanska (1969a)

Table 7 (contd).

Compound	Animal	Dose (mg/kg body weight)		Reference
		Oral	Dermal	
Maneb	mouse	4100		Engst et al. (1971)
	rat	4500		Engst et al. (1971)
	rat (male)	6750		Worthing & Walker (1983)
Nabam	rat	395		Blackwell-Smith et al. (1953)
Mancozeb	rat (female)	12 800		Ivanova-Chemi-shanska (1969b)
	rat (male)	14 000		Ivanova-Chemi-shanska (1969)
	rat	> 8000		Worthing & Walker (1983)
Propineb	rat	8500	> 1000	Worthing & Walker (1983)
	rabbit	2500		
	cat	2500		
	hen	2500		
Metiram	mouse	5400		Worthing & Walker (1983)
	rat	> 10 000		
	guinea-pig (female)	2400 - 4800		

females for 2 weeks caused severe weight loss at the highest dose level (Minor et al., 1974).

In a 2-year feeding study, 25, 250, and 2500 mg ferbam or ziram/kg diet shortened the life span of rats and caused growth depression and neurological lesions (manifested at the highest dose level by the crossing of hind legs when animals were lifted by their tail) (Hodge et al., 1956).

Groups of 24 rats (12 females and 12 males) were given a diet containing 48 mg thiram/kg diet for 2 years (a 3-generation study). No effects on growth, reproduction, blood parameters, or mortality rate were found, neither were there gross or histological changes (Van Esch, 1956). In a further study, 12 female and 12 male rats given 200 mg/kg diet for 8 months did not show any appreciable changes in growth or mortality rate, and a dose of 300 mg/kg diet for 65 weeks did not give rise to specific evidence of poisoning (Tollenaar, 1956). Groups of 24 rats (12 females and 12 males) fed diets containing 300, 1000, or 2500 mg thiram/kg diet for 65 weeks showed weakness, ataxia, various degrees of paralysis, and histological changes (calcification in the brain stem and cerebellum and dystrophic changes in the leg muscles). At 2500 mg/kg diet, there was an increased mortality rate (Fitzhugh et al., 1952). Groups of 20 young rats administered diets containing 100, 300, or 500 mg thiram/kg diet for 2 years all showed a small reduction in the growth rate. At concentrations of 300 and 500 mg/kg diet, an increased mortality rate was seen, while at 500 mg/kg diet, convulsions, thyroid hyperplasia, and calcification in the cerebellum, hypothalamus, and medulla oblongata were observed (Griepentrog, 1962; IARC, 1976).

Groups of 25 male and 25 female rats were fed diets containing 25, 250, 1250, or 2500 mg maneb/kg diet for 2 years. At 1250 mg/kg diet, there was some depression, impaired food consumption, and increased mortality rate. At the end of 2 years, the animals receiving 1250 mg/kg diet had an increased liver/body weight ratio, and those receiving 2500 mg/kg diet also showed thyroid hyperplasia and nodular goitre (Worthing & Walker, 1983).

Groups of 10 young male and 10 young female rats were fed diets containing 500, 1000, 2500, 5000, or 10 000 mg zineb/kg diet for 2 years. At the two highest dose levels, there was an apparent increase in the mortality rate among the female rats and, at 10 000 mg/kg diet, there was a tendency towards diminished growth in both sexes. The results of haematological studies were normal, but a goitrogenic effect was seen at all dose levels. Kidney damage was seen in 6 animals at the 10 000 mg/kg dose level and in one animal in each of the groups receiving 1000, 2500, or 5000 mg/kg diet, but not at all at 500 mg/kg diet. The tumour incidence was not significantly greater among any of the treated animals than it was in the controls (Blackwell-Smith et al., 1953; Kampmeier & Haag, 1954).

Weanling rats in groups of 25 males and 25 females were fed diets containing 25, 250, or 2500 mg ziram/kg diet for 2 years. The growth rate and life span were normal in all groups, but neurological changes were observed in the animals receiving 2500 mg/kg diet, though no cystic lesions were discovered in the brain. Neurological changes were not observed at lower dose levels. In some of the male animals, the testes were atrophied, and there was a slight indication of thyroid hyperplasia, notably in the 2500 mg/kg diet group. However, there was no increase in tumour incidence in the treated animals (Hodge et al., 1956). A comparable study with the same dose levels was carried out with ferbam, and again no increase in tumour incidence was found (IARC, 1976).

8.2.1.2 Dog

A dog given ferbam and ziram together for one month, each at a dose of 5 mg/kg body weight per day, remained healthy except for slight anaemia. The same result was observed when ferbam was given alone for one month at a dose of 25 mg/kg body weight per day, or for one week at 50 mg/kg body weight per day. Raising the dose to 100 mg/kg body weight per day, however, immediately provoked severe vomiting and malaise (Hodge et al., 1952). Pairs of adult dogs were given daily doses of 0.5, 5, or 25 mg ferbam/kg body weight for one year. Convulsions occurred at the highest dose level, but urine analysis, blood parameters, organ weights, and tissue histology (including that of the thyroid gland) were normal (Hodge et al., 1956).

When pairs of dogs were fed maneb orally at the rate of 2, 20, 75, or 200 mg/kg body weight per day for one year, toxic effects were observed at the two highest dose levels, but not at 20 mg/kg body weight (Worthing & Walker, 1983).

Three groups of three dogs each were fed diets containing 20, 2000, or 10 000 mg zineb/kg diet for one year. All the animals survived, and no persistent changes in growth rate were seen in any of the groups. There were no histopathological changes in the tissues, except in the thyroid gland, and haematological findings were normal. At 10 000 mg/kg diet, thyroid hyperplasia was noted (Blackwell-Smith et al., 1953; Kampmeier & Haag, 1954).

8.2.1.3 Bird

Sodium diethyldithiocarbamate (NDDC), the dimethyl compound (NDMC), and ferbam, ziram, and thiram were given orally to young and adult domestic fowl (Thorber's gog cockerels) at 330, 210, 205, 56, and 178 mg/kg body weight, respectively, and the birds were killed after 6, 12, 18, or 20 weeks. All of the compounds had an adverse effect on body weight gain, retarded testicular development, and produced degeneration in the seminiferous epithelium of mature birds. Nerve fibre degeneration was produced in the medulla and spinal cord of chicks by NDDC and in those of cocks by NDMC. Chicks exposed to thiram became lame and exhibited swollen epiphyses of the long bones due to endochondrial ossification giving rise to a thickened cartilaginous epiphyseal plate (Rasul & Howell, 1974).

8.2.2 Inhalation exposure

8.2.2.1 Rat

Studies concerning toxicity following inhalation exposure are scarce.

Ivanova-Chemishanska et al. (1972) studied the inhalation toxicity in rats with zineb (70% purity), maneb (80% purity), and mancozeb (80% purity), applied 6 days per week over a period of 4½ months, at concentrations of 2, 10, 50, 100, or 135 mg/m^3. The pesticides were given in the form of dispersed aerosols, with 95% of the dust particles ranging from 1 to 5 µm in size, and the remainder from 5 to 10 µm. Local irritation of the mucosa of the upper respiratory tract was noted and concentration-related non-specific changes in the liver and kidneys were evident. However, only slight changes were found at a concentration of 2 mg/m^3.

Davydova (1973) studied the influence of inhaled thiram on the estrous cycle and genital function of rats. Groups of rats were exposed to 0, 0.45, or 3.8 mg/m^3 thiram for 6 h/day, 5 days/week, over a period of 4½ months. An extension of the estrous cycle was seen at the highest dose level, and genital function was disturbed, as shown by a reduction in the capacity to conceive, a reduction in fertility, and of fetal weight gain.

8.3 Skin and Eye Irritation; Sensitization

Nabam (19% solution) and zineb (65% wettable powder) were each applied to the right eye of 10 rabbits, the left eye being used as a control. Nabam did not produce signs of irritation, while zineb produced mild irritation (erythema), which subsided within 6 - 8 h. No oedema was seen. The mild irritation may have been caused by the non-specific foreign body reaction to the dry, insoluble powder. When this procedure was repeated with both compounds diluted and suspended for agricultural use (for nabam, 0.5% of the commercial 19% solution plus zinc sulfate, 0.125% in water; for zineb, a 0.188% suspension of the commercial 65% wettable powder in water), no irritation was seen (Blackwell-Smith et al., 1953).

In studies performed on guinea-pigs, intracutaneous injections, 10 times daily, followed by an epicutaneous challenge test, provided evidence of the marked sensitizing and cross-sensitizing properties of thiram and metiram (Griepentrog, 1960).

It has been reported that a number of dithiocarbamates (mancozeb, metham-sodium, metiram, zineb, ziram, and thiram) cause skin and/or eye irritation (Worthing & Walker, 1983).

8.4 Reproduction, Embryotoxicity, and Teratogenicity

8.4.1 Reproduction

8.4.1.1 Rat

Groups of rats (16 male and 16 female Charles River-CD rats per group) were fed maneb for 3 months at levels of 0, 125, or 250 mg/kg diet, and were mated in a standard 3-generation, 2-litters-per-generation reproduction study. Groups of males and females from the F_{1b} and F_{2b} litters were fed maneb for 3 months after weaning and mated to become parents of the succeeding generation. The major reproduction indices were unaffected by maneb at dietary levels up to and including 250 mg/kg diet. There was no histological evidence of congenital anomalies in a variety of tissues and organs of the male and female rats of the F_{3b} litter subjected to histopathological examination (Sherman & Zapp, 1966).

Maneb, zineb, and mancozeb exert dose-dependent damaging effects on the gonads of rats of both sexes. The dose levels were 96 - 960 mg zineb/kg body weight, 140 - 1400 mg mancozeb/kg body weight, and 14 - 700 mg maneb/kg body weight, given twice a week for 4.5 months. Both reproductive and endocrine structures were affected at all dose levels, leading to decreased fertility (Ivanova-Chemishanska et al., 1973, 1975a). In a 4-month inhalation study on rats using maneb at 4.7 mg/m^3, no effect on sperm mobility was detected (Matokhnyuk, 1971).

Ivanova-Chemishanska & Antov (1980) studied the effects of Endodan^R (50% ethylenethiuram monosulfide) on the gonads and reproduction in rats during long-term daily oral doses of 3.8 or 38 mg/kg body weight. The parental generation (F_0) and 3 consecutive generations (F_1 - F_3) were examined. In F_0, a decrease in succinic dehydrogenase and ATPase activities in testes homogenates was found, as well as an increase in glucose-6-phosphate dehydrogenase (G6PDH) activity compared with control levels. Changes in the liver and brain enzyme systems were also noted.

The same results were obtained with zineb (78% purity). A rapid loss of mobility and changed resistance (to osmotic and acidic effects) of spermatozoa were found. A decreased index of fertility was also found for both sexes in the F_0 generation. Decreased index of fertility and enzymatic changes in organ homogenates were detectable in the F_1 - F_3 generations (Ivanova-Chemishanska et al., 1973).

In extracts of testes of white rats, exposed by inhalation to zineb and maneb at a concentration of 100 mg/m^3 for 4 months, Izmirova et al. (1969) found an increase in lactate dehydrogenase (LDH), LDH_2, and LDH_4. Bogartykh et al. (1979) did not find any changes in LDH or G6PDH activities in testes homogenates of Wistar rats orally treated with zineb (2.5 mg/kg body weight) for 3 months.

Thiram at doses of 225, 300, 450, 600, 900, or 1200 mg/kg diet given to male Wistar rats for 29 days produced changes in many of the parameters studied. A significant effect on testes and seminal vesicle weight was found at 450 mg/kg diet, and a decrease in body weight was found at 300 mg/kg. The most sensitive parameters studied were found to be the weights of the epididymal and perirenal fat pads, which were decreased by thiuram doses in the range 130 - 184 mg/kg diet. The no-effect level, calculated using an extrapolation model, did not differ significantly from the earlier reported value of 48 mg/kg diet (Lowy et al., 1979, 1980).

Ferbam was fed to groups of 20 Charles River-CD male rats at concentrations of 0, 500, 1200, or 2500 mg/kg diet for 13 weeks before mating with untreated females. Six of the rats fed the highest dose level died. The indices of fertility, gestation, viability, and lactation for the females mated with treated males were normal (Short et al., 1976).

When thiram was incorporated into the diet at concentrations of 0, 500, 1000, and 2500 mg/kg diet and fed to male weanling Charles River rats for 13 weeks prior to mating, food intake and growth was mainly decreased at the two highest dose levels. Loss of hair and rough coats were also seen in these groups. At the highest dose level, high mortality occurred. Males in the highest dose group failed to inseminate the females. In these animals, there was evidence of testicular hypoplasia, tubular degeneration, and atypical spermatoids in the epididymus. At the two lower dose levels, no influence on reproduction was found (Short et al., 1976).

Female rats fed 400 or 2000 mg thiram/kg diet for at least 14 days prior to mating showed a significant reduction in the number of implants per dam and pups per dam. The delaying effect on the estrous cycle was reversible. At the highest dose level, a number of animals died. A comparable study with ferbam using the same dose levels did not show any influence on fertility, gestation, viability, or lactation (Short et al., 1976).

Administration of 50 mg ziram/kg and 100 mg zineb/kg body weight to rats for a period of 2, 4, or 6 months produced delayed insemination, sterility, resorption of fetuses, and anomalies in development (Rjazanova, 1967).

8.4.1.2 Bird

Thiram (99.9% purity) has been reported to decrease egg production for the domestic chicken *(Gallus domesticus)*, pigeon *(Columba livia)*, and pheasant *(Phasianus colchicus torquatus)*. A dose level of 8.8 mg/kg body weight per day caused a 50% reduction in egg laying in bobwhite quail *(Colinus virginianus)*. During this period of reduced egg laying, it seems that an alteration of hormone levels took place resulting in significant weight losses of ovary and oviduct, decrease in serum calcium level (which is controlled by estrogen), and alteration in normal maturation of the ova (Wedig et al., 1968).

In a study by Van Steemis & Van Logten (1971), tecoram (an oxidation product of disodium EBDC and sodium dimethyl dithiocarbamate with ammonium persulfate) in propylene glycol or saline was administered to chick embryos at doses of 0.01, 0.1, 1, or 10 mg/egg. Paralysis, shortening of the extremities, muscular atrophy, dwarfing and death occurred. Microscopically, signs of peripheral neuropathy confined to the distal parts of the peripheral nerves, and muscular atrophy were found.

8.4.2 Teratogenicity

Some dithiocarbamates are potentially teratogenic in the rat, but not in the mouse. In most cases, the teratogenic effects have been observed at high dose levels.

8.4.2.1 Rat

Kaloyanova et al. (1967) have studied the effects on progeny of albino rats of 0, 700, and 1400 mg maneb/kg body weight administered twice per week for 4.5 months. Three groups of 20 rats (10 males and 10 females) were used and a first generation was bred. Congenital deformities were found in the facial part of the skull, caudal vertebrae, palates, limbs, and tail. The same type of changes were also found after a single oral dose of 2000 - 8000 mg zineb or 1000 - 4000 mg maneb/kg body weight on days 11 - 13 of pregnancy.

No teratogenic effects or adverse effects on the intrauterine development of progeny were observed when rats were given 1000 mg zineb or 500 mg maneb/kg body weight, from days 2 to 21 of pregnancy, or were exposed in an inhalation chamber to a concentration of 100 mg zineb/m^3 for 4 h/day from day 4 of pregnancy (Antonovich et al., 1972; Petrova-Vergieva & Ivanova-Chemishanska, 1973; Ivanova-Chemishanska et al., 1975a).

In a study on Sprague Dawley rats using maneb at dose levels of 0, 120, 240, or 480 mg/kg body weight on days 7 - 16 of gestation, fetotoxic effects (reduced fetal weight, reduced ossification, and hydrocephalus) were seen at the highest dose level (Chernoff et al., 1979).

In studies by Larsson et al. (1976), maneb was administered to Sprague Dawley rats at dose levels of 0, 400, 770, or 1420 mg/kg body weight, by gavage, as a single dose on day 11 of gestation. Rats were sacrificed on day 18 of gestation and fetuses were examined for reproductive and teratogenic abnormalities. A substantially increased resorption rate was seen at 770 mg/kg body weight. Gross malformations occurred in all surviving animals at 770 and 1420 mg/kg body weight, but no malformations were observed in the single litter of the low-dose group. These abnormalities included cleft palate, hydrocephaly, and other serious defects. In another study, maternal administration of zinc acetate (made in an attempt to relieve the incidence of teratogenic events) had some preventive effect at 750 mg/kg body weight, but, at 1380 mg/kg body weight, the frequency and type of malformations were unchanged (Larsson et al., 1976).

Mancozeb was administered to rats at dose levels of 0, 380, 730, or 1320 mg/kg body weight on day 11 of gestation in a study similar to that reported above with maneb. Again, a substantial increase in malformations, similar to those produced by maneb, was observed at the highest dose level, but not at lower levels (Larsson et al., 1976).

Propineb was administered to rats at dose levels of 0, 400, 760, or 2300 mg/kg body weight, by gavage, on day 11 of gestation. The dams were sacrificed and fetuses examined for gross external and internal malformations on day 18 of pregnancy. Maternal toxicity was observed at all dose levels. At the highest dose level, propineb was fetotoxic and induced a variety of malformations in the surviving fetuses. At 760 mg/kg, propineb was slightly fetotoxic but did not induce malformations in surviving fetuses. The pattern of fetal abnormalities was qualitatively similar to that noted in the maneb- and mancozeb-treated rats (Larsson et al., 1976).

Cypromate (zinc propylene bisdithiocarbamate) was studied for its teratogenic potential in white rats using either a single oral dose of 250, 500, or 1000 mg/kg body weight on the 11th or 13th day of gestation or repeated treatment from the first day of gestation through the whole pregnancy at 62, 250, and 500 mg/kg body weight. A spectrum of malformations involving the nervous and skeletal systems, facial cranium, extremities, etc., were induced with a single dose of 500 mg/kg body weight or more, and at all dose levels given repeatedly (Petrova-Vergieva, 1976).

Groups of rats (26 - 27 pregnant CD1 rats per group) were administered zineb (purity 85.5% containing 0.35% ETU) at dose levels of 0, 200, 632, or 2000 mg/kg body weight per day on days 6 - 19 of gestation. Maternal body weight and food consumption data were recorded. Pregnant rats were sacrificed at day 20 and a laparotomy was performed. Fetal data included live, dead, and resorbed fetuses as well as somatic and skeletal abnormalities. There was no maternal mortality, but a substantial weight loss was seen at the highest dose level. Fetuses from mothers administered 2000 mg/kg also showed a reduced body weight. Fetal mortality was not observed, and there were no significant anomalies noted on gross external examination. However, a higher incidence of teratogenic anomalies was noted at the highest dose level (short and kinky tails, hydrocephalus, and increased incidence of skeletal anomalies). At the 632 mg/kg level, these teratogenic anomalies were absent. The abnormalities found at the highest dose level may have been due, in part, to the presence of ETU in the formulation (Short et al., 1980).

Ferbam administered to rats on days 6 - 15 of gestation at 150 mg/kg body weight resulted in death, increased resorptions, decreased fetal weights, and a slight increase in soft and skeletal tissue anomalies (Minor et al., 1974). CD-1 rats were treated on days 6 - 15 of gestation, by gavage, with 0, 11, or 114 mg ferbam/kg body weight. Twenty-five percent of the dams administered 114 mg/kg died, but the surviving dams showed small litters, increased resorptions, and decreased fetal weight. Also, a number of malformations (unossified sternebrae, malformed cranium, hydrocephalus, and cleft palate) were found.

When thiram was administered at doses of 0, 40, 90, 136, 164, or 200 mg/kg body weight on days 6 - 15 or 7 - 12 of gestation, the 200 mg/kg dose reduced the number of mated rats that delivered litters, and only 33% of the dams survived. At doses of 136 mg/kg or more, a decrease in the number of implants and fetuses per dam, an increase in resorptions, a decrease in fetal body weight, and an increase in malformations, as described for ferbam, were observed (Short et al., 1976).

8.4.2.2 Mouse

Pregnant female NMRI and Swiss-Webster mice were treated orally during days 6 - 17 of pregnancy with thiram at 179, 357, 714, or 1071 mg/kg body weight and 250, 500, 1000, and 1500 mg/kg body weight, respectively. Increased resorption of embryos, clearly retarded fetal development, and skeletal malformation (cleft palate, wavy ribs, curved long bones of extremities, and micrognathia) were seen in both strains. The 12th and 13th days seemed to be the most sensitive period of embryonic development. The lowest dose had only a slight effect, but the next dose level was clearly teratogenic (Roll, 1971; Matthiaschk, 1973).

Thiram did not reduce body weight gain during gestation at doses of 100 or 300 mg/kg body weight, administered on days

6 - 14, and no changes in litter size, incidence of resorptions, or fetal weight were observed. However, an increase in malformations was seen (Short et al., 1976).

In studies by Larsson et al. (1976), doses of 0, 400, 770, or 1420 mg maneb/kg body weight or 0, 380, 730, or 1330 mg mancozeb/kg body weight were given on a single occasion to NMRI mice on days 9 or 13, and mice were sacrificed on day 18 of gestation. No adverse maternal or fetal effects could be detected.

In a study on CD1 mice administered 0, 375, 750, or 1500 mg maneb/kg body weight on days 7 -16 of gestation, maternal toxicity was found at the highest dose level, together with a decrease in fetal caudal ossification centres at all dose levels (Chernoff et al., 1979).

Ferbam administered to mice at 30 and 300 mg/kg body weight on days 6 - 16 of gestation did not produce any teratogenic effects (Minor et al., 1974), and, when given at 23 or 228 mg/kg body weight on days 6 - 14, it did not affect the survival or body weight of Swiss-Webster dams during gestation. No changes in litter size, incidence of resorptions, or fetal weight were observed, but, at the highest dose level, an increase in malformations was seen (Short et al., 1976).

Groups of CD-1 mice were administered zineb (85.5% purity) daily, at dose levels of 0, 200, 632, or 2000 mg/kg body weight per day, for 11 days from day 6 of gestation and sacrificed on day 18. Gross examination for maternal well-being and fetal anomalies, both somatic and skeletal, failed to show any teratogenic effects (Short et al., 1980).

8.4.3 Embryotoxicity

Korhonen et al. (1982a,b) used a system called chicken embryo test to study the embryotoxic potential of dithiocarbamates and found early and late death and malformed embryos. It was thought that this test could have a predictive value as a simple teratogenicity test, but many limitations were found in doing so, and the interpretation of the results were difficult.

8.5 Mutagenicity and Related End-Points

Seiler (1973) studied the mutagenicity of maneb and ziram. Maneb proved negative in tests with *Salmonella* strains his G46, TA1530, TA1531, TA1532, and doubtful in TA1534. Ziram was positive in TA1534, doubtful in TA1530, and negative in the other strains.

In studies by Fahrig (1974), ziram was non-mutagenic in a variety of other microorganisms (*Escherichia coli*, *Serrata marcescens*, and *Saccharomyces cerevisiae*), but Pilinskaya (1971) found that it induced chromosome breaks, most of them confined to chromosome 2, in cultured peripheral human lymphocytes.

Shirasu et al. (1976) studied mutagenicity with the rec-assay procedure, a sensitivity test using H17 rec^+ and M45 rec^- strains of *Bacillus subtilis*, and with reversion assays on plates using *E. coli* (WP2) and *S. typhimurium* TA1535, TA1536, TA1537, and TA1538. In these tests, ferbam, thiram, and ziram were non-mutagenic.

Certain dithiocarbamates given intraperitoneally to mice at 100 mg/kg body weight caused chromatid aberrations in bone marrow cells (Kurinny & Kondratenko, 1972; Hedenstedt et al., 1979), the order of effectiveness being thiram > ziram > maneb and zineb. Thiram was also shown to induce gene mutations in *Salmonella* and *Aspergillus* (Szymezyk, 1981; Zdienicka et al., 1981).

Hedenstedt et al. (1979) found that the mutagenic effect of tetramethylthiuram monosulfide (TMTM) was enhanced in the presence of metabolizing systems (S-9 mix), but that tetraethylthiuram disulfide (TETD or Antabuse^R) was not mutagenic.

Propylene bisdithiocarbamate was tested for cytogenicity by bone marrow analysis and for dominant lethal mutations by giving a single oral dose to male rats. There was a considerable increase in the number of chromosomal aberrations (chromatid fragments), reaching a maximum after 24 h (Vachkova-Petrova, 1977). In studies on the cytotoxic effects of ziram on cultures of human lymphocytes *in vitro* (Pilinskaya, 1971), the ratio of chromatid-type aberrations to chromosome-type aberrations was 2.7:1, which suggests that most chromosomal damage took place at the S-stage and the G_2 stage of the mitotic cycle. Ziram-induced chromosomal breaks were observed to be non-random, most of them occurring in chromosome 2.

Propineb and its main metabolite propylenethiourea (PTU) were investigated by the micronucleus test in mice. The following doses were given by ip injection twice, with a 24-h interval: propineb (unknown purity), 62.5, 125, or 250 mg/kg body weight in a 5% aqueous solution of Tween 80; propineb (78% purity), the same doses, but in 5% gum arabic; and PTU, 100, 200, 400, or 600 mg/kg body weight in distilled water. Controls received methanesulfonate at doses of 10, 20, 40, or 80 mg/kg (twice, in distilled water) and mitomycin at 1.75, 3.5, 7, or 14 mg/kg (twice, in distilled water). No statistically significant increase in the percentage of micronuclei was observed at any of the tested doses of propineb or PTU. The positive control groups showed the expected dose-related increase in the number of polychromatic erythrocytes with micronuclei (Rolandi et al., 1984).

Vachkova-Petrova (1981) studied the mutagenic potential of Endodan^R (ethylenethiuram monosulfide) in short-term studies. Several doses of Endodan^R were administered to groups of 6 rats either twice at an interval of 24 h or for 5 successive days, and the animals were killed 6 h after the last dose. The cells in metaphase were analysed for aneuploidy and aberrations, but no mutagenic effects that could be attributed to the chemical were detected.

Zineb and ziram were not mutagenic when tested in *Drosophila melanogaster* (Benes & Sram, 1969).

8.6 Carcinogenicity

8.6.1 Mouse

In studies by Innes et al. (1969), groups of 18 mice of each sex from two hybrid strains were given various dithiocarbamates from 7 days of age up to 18 months. The compounds were given daily, by gavage, from day 7 to weaning and thereafter added to the diet. The compounds and the respective amounts were: ferbam at 10 mg/kg body weight, then 32 mg/kg diet; maneb at 46.4 mg/kg body weight, then 158 mg/kg diet; nabam at 21.5 mg/kg body weight, then 73 mg/kg diet; thiram at 10 mg/kg body weight, then 26 mg/kg diet; and zineb at 464 mg/kg body weight, then 1298 mg/kg diet. No significant increase in tumours was found.

On the basis of all experimental data and experimental designs, IARC (1976) suggested that there was no definite proof for the carcinogenicity of maneb, though ETU, one of its metabolites, was able to produce thyroid carcinomas. However, zineb and maneb have been reported to induce pulmonary adenomas in mice when treated orally (Chernov & Khistenko, 1969; Balin, 1970).

8.6.2 Rat

A 2-year feeding study of the effects of zineb on rats was carried out using 60 young male and 60 female albino rats. These were divided into groups of 10 each and administered diets containing 0, 500, 1000, 2500, 5000, or 10 000 mg/kg diet. Growth, mortality, haematology, and organ weights were examined and histopathology was carried out. No clear influence on growth and mortality was observed, and haematological findings were within normal limits. A goitrogenic effect (hyperplasia) was observed in 50% of the animals at 500 mg/kg, and, at 1000 mg/kg diet or more, this effect was more pronounced. The interpretation of thyroid weight/body weight ratio was complicated because of the small number of animals still alive. Microscopically, no evidence of malignancies was present. At the highest dose level, kidney damage (congestion, nephritis, nephrosis) was seen. Although 10 000 mg zineb/kg diet produced moderate goitrogenic effects, this effect was also seen at

500 mg/kg in some rats, and was not clearly dose dependent (Blackwell-Smith et al., 1953).

Cases of goitre and thyroid adenoma were found in Sprague Dawley rats fed for 2 years on a diet containing 120 mg or 360 mg metiram/kg diet (Griepentrog, 1962).

8.6.3 Dog

When nine mongrel dogs (three groups of three animals each) were administered 20, 2000, or 10 000 mg zineb/kg diet for 1 year, no haematological changes were found. The thyroids of the group given the highest dose level were enlarged and showed hyper-plastic changes, but, at lower dose levels, they did not show any histological changes (Blackwell-Smith et al., 1953).

Doses of 45 mg metiram/kg body weight for 90 days or 7.5 mg/kg body weight for 23 months did not cause any ill effects (Worthing & Walker, 1983).

8.6.4 Dithiocarbamates in combination with nitrite

The above-mentioned studies were carried out with individual dithiocarbamates. Other studies have shown that these dithiocarbamates, in the presence of nitrite, can be converted to *N*-nitroso derivatives, which may be carcinogenic.

Thiram, ferbam, ziram (Eisenbrand et al., 1974; Sen et al., 1974), and disulfiram (Lijinsky et al., 1972; Elespuru & Lijinsky, 1973) react with nitrite under mildly acidic conditions to form *N*-nitroso compounds. Formation of *N*-nitrosodimethylamine (NDMA) by the action of microorganisms in sewage and soil containing 0.1% thiram has been reported to occur under experimental conditions (Ayanaba et al., 1973). Nitrite is formed by the reduction of nitrate, which can be found in some unrefrigerated vegetables (e.g., spinach and beets), especially after cooking (Phillips, 1968), in human saliva (Tannenbaum et al., 1974), and in cured meats. Thus, *in vivo* nitrosation of dithiocarbamates in the stomach cannot be totally excluded.

As has been pointed out by IARC (1976), the extrapolation of findings in experimental animals to man is complicated by many factors. It is relatively easy to show that *N*-nitroso derivatives can be formed and that these are mutagenic and/or carcinogenic. The crucial information, however, is the quantity produced in man under the prevailing conditions. The concentration of both reactants, the pH, the influence of competing reactions, and the presence of accelerators and inhibitors are all important. In addition to these difficulties in defining potential human exposure, the susceptibility of man, compared with that of experimental animals, has to be considered.

Sen et al. (1974) concluded that it is unlikely that significant amounts of NDMA would be produced from the ingestion of trace amounts of the dithiocarbamates and the normal intake of nitrite.

8.7 Mechanisms of Toxicity; Mode of Action

8.7.1 Thyroid

In weaning rats, a diet containing 500 mg nabam/kg given for 9 days caused thyroid hyperplasia and a decrease in the weight of the thymus (Seifter & Ehrich, 1948).

Male and female albino rats were administered diets containing 0, 500, 1000, 2500, 5000, or 10 000 mg zineb/kg diet for up to 30 days. Animals were killed sequentially in order to study the changes in the thyroid gland. Only in the 10 000 mg/kg group were effects seen. In two out of five males and in one out of five females, hyperplasia of the thyroid was observed (Blackwell-Smith et al., 1953). Przezdziecki et al. (1969) fed female Wistar rats a diet containing 1300 mg zineb/kg or 1875 mg maneb/kg diet for 7 months. Significant increases in the weight of the thyroid gland and decreases in the weight of the kidneys, adrenal glands, and ovaries were observed.

Thyroid hyperplasia has been reported in rats given maneb, zineb, or mancozeb in amounts ranging from 500 to 2500 mg/kg diet for periods of up to 2 years (FAO/WHO, 1965b, 1971b). In a 2-year feeding study, 2500 mg maneb/kg diet produced thyroid hyperplasia and nodular goitre and increased mortality, but 1250 and 250 mg/kg diet did not cause any ill effects (FAO/WHO, 1965b).

Zineb was given orally to white rats at dose levels of 96 or 960 mg/kg body weight for 4.5 months. Compared with that of untreated animals, the thyroid was enlarged with microfollicles and columnar cells. Succinic dehydrogenase and cytochrome oxidase activities were raised in these cells, while the colloid in the follicles showed reduced PAS[a]-positive granules. These changes were consistent with an increase in thyroid-stimulating hormone (TSH). An increased number of basophilic cells containing PAS[a]-positive granules was observed in the adenohypophysis (anterior pituitary). These effects were seen only at the highest dose level. The uptake of ^{131}iodine was also increased at the highest dose level, and a high plasma TSH level was recorded in treated animals. The changes observed in both thyroid and pituitary were probably a compensatory response to the antithyroid effect of the dithiocarbamate (Ivanova-Chemishanska et al., 1975b).

---

a PAS = periodic acid-Schift reagent.

After oral administration of zineb to rats at dose levels of 9.6 or 960 mg/kg body weight, twice a week, for 4½ months, the gonadotropic and thyroid-stimulating functions of the adenohypophysis were significantly increased compared with those of control values, more markedly in those receiving the higher dose (Ivanova-Chemishanska et al., 1974).

Albino rats were orally administered doses of 2400 mg zineb/kg or 3500 mg maneb/kg body weight, and, after 24 h, radioactive iodine was administered intraperitoneally. Reduced assimilation of ^{131}iodine by the thyroid was found, which suggests that the dithiocarbamates, or certain of their metabolites, possess a marked antithyroid effect and inhibit the synthesis of thyroxine (Ivanova-Chemishanska et al., 1967, 1974). In similar studies with mancozeb, a single oral dose of 7000 mg/kg body weight resulted in a decreased uptake of ^{131}iodine (Ivanova-Chemishanska et al., 1967).

The condition of the thyroid gland was studied in male albino rats, which were administered 700 mg maneb/kg, 960 mg zineb/kg, or 1400 mg mancozeb/kg body weight. After 30 days, the distinct morphological changes observed indicated stimulation of the thyroid by TSH. Hypophyseal stimulation is the consequence of release from negative feedback by thyroxine, the plasma level of which is depressed by the action of EBDCs (Ivanova-Chemishanska et al., 1971, 1974).

Rats fed a diet containing 10 000 mg metiram/kg for 2 weeks showed increased thyroid weight and a decrease in the uptake of ^{131}iodine, but no ill effects were produced by 1000 mg/kg diet (Worthing & Walker, 1983).

8.7.2 Interaction of dithiocarbamates and alcohol

Hald et al. (1948) found that dithiocarbamates interact with ethanol (ethyl alcohol), and since then, certain dithiocarbamates, particularly disulfiram, have been used in the treatment of chronic alcoholism. Disulfiram has been proposed to act in two different ways. The first possibility is that the drug or one of its metabolites (e.g., diethyldithiocarbamate, carbon disulfide) interferes with the normal metabolism of ethanol and, consequently, gives rise to an accumulation of toxic amounts of intermediary products, such as acetaldehyde. The second possible method of action is that ethanol interferes with the normal metabolism of disulfiram and therefore makes disulfiram more toxic in some way.

Ethanol is detoxified in many tissues, particularly the liver, by oxidation, firstly to acetaldehyde, then to acetic acid, and finally to carbon dioxide and water. Disulfiram interferes with various enzyme systems including those involved in the oxidation of ethanol. After administration of disulfiram, the blood acetaldehyde level increases significantly.

Peripheral neuropathy and optic neuritis have been observed in alcoholics treated with 125 - 150 mg disulfiram per day (Gardner-Thorpe & Benjamin, 1971).

Van Logten (1972) studied this phenomenon of alcohol intolerance extensively in rats. Zineb and maneb did not induce alcohol intolerance, whereas most of the alkyldithiocarbamates (such as ziram and nabam) and thiuram sulfides (such as thiram) did. In general, the dithiocarbamates with a free H atom bound to the N atom did not induce intolerance. Apart from the accumulation of acetaldehyde in blood, disturbance of sulfobromophthalein (BSP) elimination, increased serum glutamic-pyruvic transaminase, hypothermia, increased glucose content, changes in blood morphology, atrophy of spleen and thymus, and increase in the weight of adrenals and brain were all observed. Studies on adrenalectomized rats showed the involvement of the adrenals in the alcohol intolerance. Hyperglycaemia, eosinopenia, lymphopenia, and neutrophilia were not seen in the animals without adrenals, and spleen and thymus atrophy was reduced. However, the effect on the red blood cells was more pronounced, and the accumulation of acetaldehyde in the blood was unaffected by adrenalectomy.

Oral treatment of rats with ethanol after administration of alkyldithiocarbamates or thiuram sulfides lowers the catecholamine content of the adrenals. Since the lowest level was not reached until 24 h or more after alcohol treatment, it seems likely that the influence on the adrenals of the dithiocarbamate-ethanol interaction is of secondary importance. There was no clear indication that the ethanol intolerance in the rat is accompanied by changes in brain catecholamine level.

As in human beings, the dithiocarbamate-ethanol reaction in the rat is characterized by a severe hypotension, which starts almost immediately after the administration of the ethanol and lasts at least 8 h. It is evident that, during alcohol intolerance, body fluids shift from the plasma into the interstitial tissue or cells, possibly thereby causing the hypotension or shock. The observed hypothermia also starts almost immediately after ethanol administration and may last for many hours. Adrenalectomy did not prevent the hypothermia. Therefore, this phenomenon must be considered as a primary effect, along with the plasma accumulation of acetaldehyde.

Since heat loss is not increased during the dithiocarbamate-ethanol reaction, the hypothermia is probably due to decreased heat production. However, the serotonin concentration in the brain was increased, and so a disturbance of the thermoregulation cannot be ruled out. It seems doubtful that the dithiocarbamate-ethanol reaction is due to acetaldehyde *per se*. Intraperitoneal injection of acetaldehyde, which resulted in a blood level twice as high as during the dithiocarbamate-ethanol reaction, did not influence BSP elimination, serum glucose level, body temperature, organ weights, catecholamine content of the adrenals, or blood pressure.

No indications were available that the accumulation of acetaldehyde in the blood was due to an accelerated biotransformation of alcohol. More work needs to be done to decide whether the accumulation of acetaldehyde and pyruvate in the blood is a consequence of a disturbance of carbohydrate metabolism. The specificity of ethanol is remarkable. The combination of thiram with either methanol or 1-propanol has no effect on BSP elimination or on blood glucose level.

The sensitivity of the rat for the dithiocarbamate-alcohol reaction is of the same magnitude as that of man. Administration of 1.9 mg thiram/kg body weight in the rat elicits an accumulation of acetaldehyde in the blood. Thus, it may be concluded that 60 ml of gin, 0.5 litre of beer, or even less should be sufficient to induce alcohol intolerance.

Eight days after the administration of certain dithiocarbamates, a dithiocarbamate-alcohol reaction may be observed when ethanol is given. The maximum level of acetaldehyde in the blood is reached within 15 min of ethanol treatment.

A 90-day study with several dietary levels of thiram revealed a no-toxic-effect level of 100 mg/kg diet. After feeding rats with 10 mg thiram/kg diet for 6 weeks, oral administration of a single dose of 6 ml ethanol/kg body weight caused a significant decrease in the body temperature. Higher doses of thiram with alcohol induced hyperglycaemia, accumulation of acetaldehyde in the blood, and other abnormalities. In contrast, the combination of 100 mg thiram/kg diet and 5% ethanol continuously in the drinking-water did not have any effect (Van Logten, 1972).

8.7.3 Neurotoxicity

In an 80-week study on the neurotoxic and behavioural effects of thiram, 12 male and 12 female rats per group were fed thiram at dose levels of 0, 100, 400, or 1000 mg/kg diet (the concentration of the compound in the diet was periodically increased in order to give a relatively constant consumption on the basis of body weight). A second study was carried out on two groups of 24 female rats administered 0 or 1000 mg thiram/kg diet, for 36 weeks. The neurotoxic effects were characterized by ataxia and paralysis of the hind legs, although these effects were only seen at the highest dose level (1000 mg/kg diet, equivalent to 65 mg/kg body weight) in females. Demyelination, degeneration of the axon cylinders, and the presence of macrophages in the nerve bundle of the sciatic nerve were seen. Degeneration in the ventral horn of the lower lumbar region of the spinal cord was demonstrated by chromatolysis of motor neurons, pyknosis, and satellitosis. Electromyograms indicated a loss of motor unit function, and the histopathology suggested that the peripheral nerve is the primary site of the lesion (Lee & Peters, 1976).

In another study, groups of 12 males and 12 females were fed ferbam in the dose levels that gave actual intake levels of approximately 8.5, 34, and 87 mg/kg body weight (average of males and females) per day. The neurotoxic effects of ferbam are less than those of thiram. In this study, only 3 of the 24 rats fed the highest dose level developed ataxia or paralysis (Lee & Peters, 1976). Neurotoxic effects have also been observed for ziram by Hodge et al. (1956).

In a study on rats, zineb (490 and 2450 mg/kg body weight), maneb (350 and 1750 mg/kg body weight), and mancozeb (700 and 3500 mg/kg body weight) were administered orally at twice weekly doses for 4 months. Mortality was high, and paresis in the hind limbs appeared in the third month of the study and progressed to complete paralysis (Ivanova-Chemishanska, 1969a).

Dishovski & Ivanova-Chemishanska (1979) studied the ultrastructural changes in the neocortex of rats repeatedly administered propineb (70% purity) at 85 mg/kg and 425 mg/kg body weight for 40 days. At the higher dose level, intense ultrastructural changes in the sensorimotor neocortex were detected using an electron microscope, primarily affecting the pyramidal cells. The concentration of ribosomes and hypertrophy of the Golgi apparatus suggested an increase in synthetic processes in the neurons.

Edington & Howell (1966, 1969) found lesions in the central nervous system of adult Dutch-New Zealand rabbits who were given ip injections of sodium diethyldithiocarbamate (NDDC) at 330 mg/kg body weight, for 6 days/week, for 30 weeks. The first changes were seen at 6 weeks in the accessory cuneate nucleus and in Clarke's column; 12 weeks later, degeneration was seen in the spinocerebellar tracts in the cerebellum medulla. After 24 weeks, severe nerve fibre degeneration in the peripheral white matter of the spinal cord (both involved the axon and myelin sheath) was observed. It was suggested that these changes might be connected with changes in the level of copper in the serum.

Kim & Rizzuto (1975) studied the effect of NDDC (0.23, 2.3, and 23 μg/ml nutrient medium) on myelinated cultures of new-born mouse cerebellum. Exposure time was 24 - 120 h, and the cultures were examined by light and electron microscopy. Treatment of the cultures for 24 - 48 h produced swelling of axons and presynaptic endings, morphologically characteristic of dystrophic axons. Continued exposure induced an extensive degeneration of axons and myelin sheath (Wallerian degeneration in axons).

8.7.4 Dithiocarbamates in combination with metals

Truhaut et al. (1971) studied the chelating action of sodium diethyldithiocarbamate to copper and the fact that this element is indispensable for the activity of dopamine β-hydroxylase. The authors put forward the hypothesis that the inhibition of this enzyme system, which catalyses the conversion of dopamine to norepinephrine and participates in the biogenesis of catecholamines in the central nervous system, may play a role in the etiology of neurotoxic effects.

Maj et al. (1970) studied the effect of disulfiram, diethyldithiocarbamate (DDC), and dimethyldithiocarbamate on serotonin (5-HT) and 5-hydroxyindole-3-acetic acid (5-HIAA) in the brain of rats. The total dose levels ranged from 150 to 500 mg/kg body weight. It was concluded that these three compounds do not affect the 5-HT level in the rat brain. The 5-HIAA levels increased, but not significantly.

Possibly, reactions of carbon disulfide with pyridoxamine could lead to the depletion of pyridoxal phosphate in the tissues, which may, in turn, cause neurological changes. Long-term poisoning of rabbits with carbon disulfide has been shown to result in increased excretion of zinc in the urine and disturbances of copper and zinc concentrations in the tissues. Also, after NDDC treatment, increased levels of copper in the liver and nervous tissue have been found (Cavanagh, 1973).

Aaseth et al. (1981) showed that oral treatment of Wistar rats with tetramethylthiuram disulfide (TMTD) at 1000 mg/kg diet for one week increased the brain levels of endogenous copper and zinc. In further studies, rats were administered an iv injection of $^{203}HgCl_2$ (5 μmol/kg body weight in saline) at day 17 of pregnancy. DDC was given immediately after the mercury injection (500 μmol/kg body weight). The maternal brain concentration of mercury increased significantly, and the kidney levels, measured after 24 and 48 h, also increased. In the fetuses, the mercury in the brain, liver, kidneys, and blood (but also in the placenta) were significantly increased after 24 h, but, after 72 h, only the levels in fetal blood were still elevated. Mice of the NMRI strain were similarly injected with $^{203}HgCl_2$ (2.5 μmol/kg body weight) and fed diets containing DDC (10 000 mg/kg diet), disulfiram and TMTD (1000 mg/kg diet), or carbon disulfide (3000 mg/kg diet) for 4 days. The brain level of mercury was significantly increased after DDC or TMTD treatment and marginally after disulfiram or carbon disulfide treatment (Aaseth et al., 1981).

Lakomaa et al. (1982) studied the effect of DDC on copper and zinc concentrations in different regions of the brain of Long-Evans rats during acute or repeated treatment. Acute treatment (250 mg/kg body weight) produced no effect after 24 h, whereas repeated treatment (250 mg/kg, 5 times per week, for

4 weeks) increased copper levels in the brain stem, cortex, hippocampus, and the rest of the brain, but did not alter zinc concentrations.

Dithiocarbamates, with their metal-chelating properties, and thiuram derivatives, have been demonstrated to cause a marked increase in the concentration of lead in the brain as well as a redistribution of lead in the rest of the body (Oskarsson, 1983, 1984; Danielsson et al., 1984). Thus, after injection of a dose of labelled lead (^{203}Pb), the brain concentrations were increased by up to 100 times in thiuram-treated rats.

Male Sprague Dawley rats (10 groups of 5 rats each) were administered different combinations of thiram, disulfiram, DDC, or dimethyldithiocarbamate in combination with sodium or lead. The study demonstrated that treatment with dithiocarbamates and thiram derivatives in rats exposed for 6 weeks to lead causes a substantial (up to 4-fold) increase in the lead concentration of the brain. This effect can be explained by the formation of a lipophilic lead-dithiocarbamate complex, which probably is retained longer and has a higher capacity to penetrate the blood-brain barrier and bind to lipid-rich brain tissue components than inorganic lead itself. The chemical form of the lead when it is in the brain remains uncertain. The lead complex may decompose in the brain into inorganic lead, which exerts a neurotoxic effect, or it may be very stable in the brain and of low toxicity for the central nervous system (Oskarsson & Lind, 1985).

There are several reports on the effect of dithiocarbamates on the distribution in the body of other metal ions such as cadmium, thallium, nickel, copper, zinc, and mercury (Oskarsson & Lind, 1985).

DDC has been shown to have a strong inducing effect on levels of metallothionein, a low molecular weight, heavy-metal-binding protein, in rat liver and kidney. The mechanism probably reflects enhanced uptake of copper and depletion of hepatic glutathione (Sunderman & Fraser, 1983).

8.7.5 Miscellaneous reactions

Dithiocarbamates, with their chelating capacity, also interfere with a number of enzyme systems containing metals such as zinc and copper (e.g., dopamine β-hydroxylase). They also inhibit sulfhydryl (SH)-containing enzymes and a number of other enzyme systems involved in glucose metabolism (e.g., hexokinase, glyceraldehyde-3-phosphate dehydrogenase, and glucose-6-phosphate dehydrogenase). The effect of dithiocarbamates on liver enzymes has consequences for the metabolism of other chemicals. Thus, the toxicity of carbon tetrachloride is decreased by diethyldithiocarbamate (Lange & Jung, 1971; Lutz et al., 1973), and the toxicity of other chemicals, e.g., ethyl alcohol, may be increased (section 8.7.2).

9. EFFECTS ON MAN

9.1 Occupational Exposure

9.1.1 Acute toxicity - poisoning incidents

The acute toxicity of dithiocarbamates is low and, therefore, acute intoxication in human beings is unlikely to occur.

A case was reported of a 62-year-old man with acute kidney insufficiency after maneb application. However, the precise cause of maneb exposure was not clear, since the patient had a history of hypertension, cerebral infarction, gastrectomy because of stomach cancer, and chemotherapy. The patient was treated with haemodialysis and was discharged from hospital (Koizumi et al., 1979).

Thiram (100 mg/m^3) has been shown to cause headaches, vertigo, impairment of mental capacity, muscle twitch, and paraesthaesia (Sprecher & Grigorowa, 1967).

9.1.2 Case reports, short-term and epidemiological studies

9.1.2.1 Dermal

The irritant and allergic potential of most dithiocarbamates is evident in occupational exposure. Skin irritation and sensitization were studied in man using a conventional patch test. A cotton square was dipped in 19% nabam solution and placed on the inner surface of the forearm, and, 14 days later, this procedure was repeated on the opposite forearm. Zineb was tested in the same manner, except that the cotton square was dipped in 65% wettable powder. The patches were left in place for 48 h. Of the 25 subjects included in the nabam study, 2 showed irritation (mild erythema and itching). Thirteen of the 25 reacted to the retest (from mild erythema to severe erythema, oedema, and vesiculation), indicating sensitization. Of the 50 subjects used in the zineb study, no reaction at all was seen in 49 of them. One reacted in such a way that it indicated primary irritation rather than sensitization (Blackwell-Smith et al., 1953). Schultheiss (1957) reported a case of contact dermatitis with thiram. Zadorozhny et al. (1981) found dermatitis and eczema in 241 industrial workers exposed to TMTD and other types of pesticides. Twenty-one of them showed contact dermatitis, 25 allergic dermatitis, and 7 eczema.

Cases of diffuse erythema and eczematoid epidermatitis of the eyelids and inguinal regions, probably with elements of sun sensitization, were observed among agricultural workers (grape and tobacco industries) in contact with zineb (Babini, 1966) or maneb (Laborie & Laborie, 1966; Zorin, 1970). These were largely allergic in character with only a few manifestations of contact dermatitis. Decreased resistance of the workers, vitamin deficiency, chronic liver disease, and other factors apparently contributed to these effects.

9.1.2.2 Exposure via different routes

Kaskevich et al. (1981) carried out an epidemiological study on 137 workers engaged in zineb manufacturing (51 men and 86 women). The duration of exposure to zineb for 52 workers was between 1 and 3 years, and for 85 workers between 4 and 5 years. Control groups in this study consisted of 193 persons, not exposed to chemicals and matched for age, period of employment rate, and sex. The concentrations in the air of the working area never exceeded 1 mg/m^3. Among workers occupationally exposed to zineb, the following changes were found: hepato-cholecystitis (28.4% of workers, versus 13.5% in controls); vegetovascular dystonia connected with disorders in the central nervous system (34.9%, versus 22.3% in controls); chronic bronchitis (4.4%, versus 0.5% in controls); contact dermatitis (11.9%, versus 0.1% in controls); and disorders in the menstrual cycle (16.91%, versus 4.3% in controls). These studies indicate a change in catecholamine metabolism.

In a study with cultured lymphocytes from 15 workers working in different stages of zineb manufacture, the mean incidence of aberrant metaphases was 6% greater than that in controls. The incidence of chromosomal aberrations (chromatid breaks) in cultured human lymphocytes treated with maneb (0.5, 15, or 30 μg/ml) was 10 - 20% greater than in controls (Antonovich et al., 1972).

A number of studies on maneb and mancozeb production workers have been carried out. In the earliest study (1965), 54 production workers were given medical examinations that included blood and urine analyses. Since this study predated the availability of immunoassay techniques for thyroid hormone determination, protein-bound iodine was used as a measure of thyroid function. No thyroid or other medical abnormalities could be attributed to EBDC exposure. In a second study (1975), 57 exposed and 98 unexposed production workers were examined for thyroid function by measuring triiodothyronin (T_3), thyroxine (T_4), and TSH. Again, no effects attributable to work-place exposure were identified. Workers exposed to EBDC levels ranging from 0.13 to 5.46 mg/m^3 were found to have elevated ETU and manganese levels in the urine. In a 1976 mortality study, 992 past and present production workers (over the period 1948-75) were studied. Compared with the local general population, neither the overall death rate nor the death rate due to cancer was elevated. The number of cancer deaths observed (10) was too small to evaluate cancer-specific mortality (Gowers & Gordon, 1980).

In the most extensive study, 42 currently exposed and 112 previously exposed workers were compared with equal size control groups matched for age, period of employment, race, and type of job. All participants were given thorough physical examinations by specialists in diagnostic medicine, including detailed questionnaires and interviews about health history and family health. A separate thyroid examination was carried out by thyroid specialists. Thyroid parameters that were measured included total T_3, T_3 resin uptake, T_4, TSH, free T_4 index, thyroglobulin antibodies, and microsomal antibodies. In addition, urine was analysed for ETU, EBDC, zinc, manganese, creatine, iodide, specific gravity, and pH. Blood levels of glucose, urea nitrogen, sodium, potassium, calcium, chloride, carbon dioxide, cholesterol, total protein, protein albumin, bilirubin, uric acid, creatinine, inorganic phosphate, lactic dehydrogenase, and serum glutamic oxaloacetic transaminase were also determined. As in earlier studies, the occurrence of unusually high levels of ETU in the urine of currently exposed workers confirmed their exposure. However, a detailed statistical analysis of the data revealed no differences in thyroid function, blood and urine indicators of liver and kidney function, or general health, between exposed and control groups (Gowers & Gordon, 1980; Charkes et al., 1985).

PART B

ETHYLENETHIOUREA (ETU) AND PROPYLENETHIOUREA (PTU)

CONTENTS

INTRODUCTION

One of the metabolic products of ethylene bisdithiocarbamate decomposition in mammals, plants, and lower organisms is ethylenethiourea (ETU). It may also be present as an impurity in these dithiocarbamates, and their residues on crops may be partly transformed into ETU during food processing. A comparable breakdown takes place with propineb, giving rise to propylene-thiourea (PTU).

1. SUMMARY

1.1 Sources, Environmental Transport, and Distribution

Ethylenethiourea (ETU) is found together with residues of the parent ethylene bisdithiocarbamates (EBDCs) in and on crops that have been treated with these pesticides. During storage, processing, and cooking, the amount of the parent compound decreases while that of ETU increases. ETU is easily photo-oxidized (in the presence of photosensitizers) to ethyleneurea (EU).

1.2 Environmental Levels and Human Exposure

In certain crops, such as spinach, carrots, and potatoes, treated with EBDCs, high levels of ETU can be found after cooking. In general, however, the ETU levels are below 0.1 mg/kg product.

Estimates of the exposure of the general population of the USA are of the order of 0.24 - 3.65 μg ETU/kg body weight per day, and, in Canada, estimates based on market-basket studies are around 1 μg ETU/kg body weight per day.

1.3 Kinetics and Metabolism

ETU is rapidly absorbed, metabolized, and excreted in mammals. Up to 90% is eliminated via the urine and only a small amount via the faeces. Distribution of ETU in the body appears to be fairly uniform with the exception of a relative accumulation in the thyroid. ETU is broken down to ethylene diamine (EDA), urea, carbon dioxide, or oxalic acid, or is transformed to imidazole derivatives in mammals, plants, and the environment.

1.4 Effects on Organisms in the Environment

The available LC_{50} levels of ETU and EU for fish are in the range of 7500 - 13 000 mg/litre.

1.5 Effects on Experimental Animals and *In Vitro* Test Systems

1.5.1 Ethylenethiourea

The acute oral toxicity in experimental animals is low, and the long-term effects are mainly characterized by an antithyroid action.

At dose levels > 25 mg/kg body weight, decreases in serum T_3, T_4, and protein-bound iodine (PBI) and increases in thyroid-stimulating hormone (TSH) have been found in studies on experimental animals. At higher dose levels (> 100 mg/kg body weight), increases in thyroid weight and hyperplasia occurred, which finally resulted in the development of adenocarcinoma. The effects of short-term exposure to low levels of ETU seem to be reversible, but those of long-term exposure to higher levels become, at a certain stage, irreversible. A level of approximately 5 mg/kg body weight seems to be without effects.

Most mutagenicity studies on ETU, especially those with mammalian test systems, have given negative results.

A number of carcinogenicity studies have been carried out on mice, rats, and hamsters. In addition to an antithyroid action, ETU has been found to induce, subsequently, thyroid tumours (hyperplastic goitre, solid-cell adenomas, and follicular and papillary carcinomas) in mice and rats. In an earlier study on mice, liver tumours, lung tumours, and lymphomas were also detected, but these findings have not been confirmed. No tumours except thyroid tumours have been found in rats, and in hamsters, no tumours of the thyroid gland or other organs were observed, even at 200 mg/kg diet.

At dose levels above approximately 10 mg/kg body weight, ETU has clear teratogenic effects in rats and hamsters, different types of central nervous system and skeletal anomalies being induced. However, in mice, no teratogenic effects were found at much higher dose levels (up to 800 mg/kg body weight).

1.5.2 Propylenethiourea

In a long-term study on mice using propylenethiourea (PTU), an increased incidence of hepatocellular adenomas was observed at dose levels of 10 mg/kg diet or more. No thyroid tumours were found, but increased thyroid hypercellularity occurred at a dose level of 1000 mg/kg diet. In rats, goitrogenic effects were seen with PTU at dose levels as low as 1 mg/kg diet.

1.6 Effects on Man

Epidemiological studies on workers exposed to ETU did not reveal any increase in the incidence of thyroid tumours.

2. IDENTITY, PHYSICAL AND CHEMICAL PROPERTIES, ANALYTICAL METHODS

2.1 Identity

The chemical structure of ethylenethiourea (ETU) is:

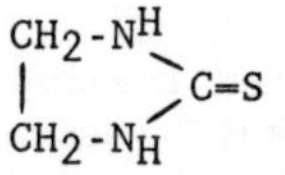

2.2 Physical and Chemical Properties

ETU is a fairly stable compound. Some physical and chemical properties are listed in Table 8.

Table 8. Some physical and chemical properties of ETU

Empirical formula	$C_3H_6N_2S$
Common synonym	2-imidazolidinethione
Appearance	white, crystalline
Relative molecular mass	102.17
Odour	odourless
Melting point	203 - 204 °C
Solubility	in water: 20 000 mg/litre at 30 °C; in ethanol: moderately soluble; in chloroform: nearly insoluble[a]
CAS registry number	96-45-7

[a] From: IARC (1974), IUPAC (1977), US EPA (1984).

2.3 Analytical Methods

Residue analysis consists of sampling the contaminated material, extracting the pesticide residue, cleaning up the extract of interfering substances, and identifying and quantifying the pesticide residue. The main methods used are summarized in Table 9.

2.3.1 Extraction

Methanol and ethanol have been used as extraction solvents for biological samples, due to the high solubility of ETU in polar solvents. Mixed solvents such as methanol/chloroform (Onley & Yip, 1971) or methanol/acetone (Phillips et al., 1977) have also been employed, and the addition of trichloroacetic acid has been reported to improve recovery with the latter solvent. Sodium ascorbate has also been found effective in ensuring good recovery of ETU (IUPAC, 1977; Otto et al., 1977).

2.3.2 Clean-up

The simplest procedures involve extraction of a derivative from aqueous acid and alkali (Newsome, 1972; Nash, 1974; King, 1977). Another approach has been to purify the initial extract by column chromatography before proceeding with derivatization (Onley & Yip, 1971; Haines & Adler, 1973; Onley, 1977) or determination steps (Otto et al., 1977). Where ETU is determined without derivatization, a solvent-partitioning step is included to provide further clean-up (IUPAC, 1977; Otto et al., 1977).

2.3.3 Derivatization

In all cases, derivatization involves first an alkylation of the thiocarbonyl group. The various derivatives that have been used are given in Table 9. Careful attention to reagent purity is essential to ensure quantitative results (Onley & Yip, 1971; Pecka et al., 1975; King, 1977). The benzyl chlorides react smoothly by refluxing in alcohol for 30 min, while alkylation with butyl bromide is carried out at room temperature in aqueous dimethylformamide containing sodium hydroxide and sodium borohydride. Solutions of ETU in aqueous dimethylformamide have been found to be extremely unstable and must be reacted immediately (Phillips et al., 1977). The *n*-butyl (Onley & Yip, 1971) and *m*-trifluoromethyl benzyl (King, 1977) derivatives are sufficiently volatile to be analysed directly by gas-liquid chromatography, whereas the benzyl derivatives must be concentrated and acetylated before quantifying. Care must be exercised during the concentration step to prevent losses through evaporation (Pecka et al., 1975). Pentafluorobenzoyl chloride (Nash, 1974) and trifluoroacetic anhydride have been used as acetylating reagents, the former requiring a column chromatographic step to remove excess reagent and by-products before moving to gas-liquid chromatography. Although the excess trifluoroacetic anhydride is easily removed by evaporation, the trifluoroacetate derivative is unstable in the presence of moisture and must be determined soon after removal of the excess reagent (IUPAC, 1977).

2.3.4 Determination

Gas-liquid chromatographic methods predominate because of their greater sensitivity, specificity, and accuracy. Methods of determining ETU in plant samples were reviewed in 1976 by the IUPAC Commission on Pesticide Terminal Residues (IUPAC, 1977), and an extensive review of methods for ETU determination has been produced by Bottomley et al. (1985).

2.3.4.1 Gas-liquid chromatography (GLC)

A variety of column packings and conditions have been used in the determination of ETU and its derivatives. Detectors used include thermionic (Onley & Yip, 1971), flame photometric (FPD) (Haines & Adler, 1973; Onley, 1977; Otto et al., 1977), and electron capture (EC) (Nash, 1974; King, 1977). Although quantification by GLC/EC enables the use of smaller samples (5 - 10 g) for monitoring ETU residues at the 0.01 mg/kg level, it requires confirmation of suspected residues by mass spectrometry (MS), a second derivative, or by element-selective detectors. Methods employing GLC/FPD with large samples (40 - 100 g) have the advantage of both quantifying and confirming ETU residues.

2.3.4.2 Thin-layer chromatography (TLC)

A variety of adsorbents and developing solvents have been used to detect ETU in plants (Vonk & Kaars Sijpesteijn, 1970; Onley & Yip, 1971; Blazquez, 1973; Engst & Schnaak, 1974). The limit of detection is 0.02 mg/kg using alumina plates and Grotes reagent for visualization (Onley & Yip, 1971). Semi-quantitative determinations are possible by comparison with ETU standards run simultaneously (IUPAC, 1977).

2.3.4.3 Polarography

This technique involves clean-up on an alumina column, followed by paper chromatography and determination of the nitroso derivative by polarography (Engst & Schnaak, 1974).

2.3.4.4 Radioisotope dilution

A reverse isotope dilution method has been used to determine ETU in the presence of its metabolites, and is useful in the low milligram range (Graham & Bornak, 1973; IUPAC, 1977).

2.3.4.5 High-pressure liquid chromatography (HPLC)

High-pressure liquid chromatography has been used for the determination of ETU without derivatization. Detection can be by ultraviolet absorption or electro-conductivity measurement, the minimum level being 0.025 mg ETU/litre or kg (Prince, 1985). Massey et al. (1982) reported an HPLC method applied for ETU determination in a beer extract with a detection limit of 10 μg/kg. The method has been found to give spuriously high results in the determination of ETU in beer due to the presence of co-eluting matrix components. The more powerful resolving ability of column-switching high-performance liquid chromatography, using polar-bonded columns of different selectivities, has proved highly effective in separating ETU from these co-eluting materials.

Table 9. Methods for the determination of ETU in plant samples[a]

Extraction solvent	Extraction/clean-up	Derivative formation	Analysis measurement[b]	Detectability (mg/kg)	Reference
Plant extract	ethanol	none	silica gel\TLC	10.0	Vonk & Kaars Si pesteijn (1970)
Plant extract	ethanol	none	paper electrophoresis	-	Vonk & Kaars Si pesteijn (1971)
Ethanol and chloroform	cellulose column	2-(butylthio) - 2-imidazoline	GLC\thermionic detector	0.02	Onley & Yip (1971)
Methanol	cellulose column		GLC/FPD	0.002	Watts et al. (1974)
Methanol	chloroform/HCl		GLC/ECD	0.005	Newsome (1972)
Methanol	Al_2O_3 column	2-(butylthio)-2-imidazoline	GLC/FPD	0.05	Haines & Adler (1973); Onley (1977)
Dioxane and water	none	none	silica gel/TLC	-	Blazquez (1973)
Methanol/ Na-ascorbate	Al_2O_3 column	none	GLC/FPD	0.01	Otto et al. (1977)

Table 9 (contd).

Extraction solvent	Extraction/clean-up	Derivative formation	Analysis measurement[b]	Detectability (mg/kg)	Reference
Methanol	florisil column	2-(benzylthio)-1-(pentafluorobenzoyl)-2-imidazoline	GLC/ECD	0.005	Nash (1974)
Ethanol	ether/HCl partition	2-(m-trifluoromethylbenzylthio)-ETU	GLC/ECD	0.01	King (1977)
Acetone	methanol/acetone; Al_2O_3 column acetonitrile/dichloromethane silica column	2-(benzylthio)-1-pentafluorobenzyl)-2-imidazoline	GLC/ECD	0.01	Newsome (1978)

a From: IUPAC (1977).

b TLC = thin-layer chromatography; GLC = gas-liquid chromatography; FPD = flame photometric detection; ECD = electron capture detection.

3. SOURCES OF HUMAN AND ENVIRONMENTAL EXPOSURE

During recent years, much attention has been paid to the finding that ETU may occur in plant samples following the use of dithiocarbamate fungicides. It may be present in the fungicide when applied, or may result from subsequent transformation (Bontoyan et al., 1972). Similarly, propylenethiourea (PTU) may occur in residues of the fungicide propineb (IUPAC, 1977). The amounts of ETU present in commercial formulations vary from one sample to another, and depend on the length of time between manufacture and use and the storage conditions, especially temperature and moisture. Bontoyan & Looker (1973) found that ETU increased, during storage for 39 days at 49 °C and 80% relative humidity, from an initial content of 0.02 - 2% to a final level of 0.13 - 14.5%. The degradation dynamics of formulations from different manufacturers varied, products containing both manganese and zinc forming the least ETU (IUPAC, 1977).

ETU is one of the important residues in plants and in the environment following the agricultural use of ethylene bisdithiocarbamates (EBDCs). It is also a metabolite formed when EBDCs are ingested by animals and man.

Sources of human and environmental exposure to ETU are also discussed in sections 3, 4, and 5 of Part A and sections 4 and 5 of Part B.

4. ENVIRONMENTAL TRANSPORT, DISTRIBUTION, AND TRANSFORMATION

ETU is a fairly stable compound with respect to hydrolytic reactions but is easily oxidized to ethyleneurea (EU). Oxidation to EU takes place primarily in biological systems and by photolytic reaction, especially in the presence of photosensitizers (Cruickshank & Jarrow, 1973; Ross & Crosby, 1973). In studies by Kaars Sijpesteijn & Vonk (1970), pure cultures of soil bacteria and fungi were unable to effect this transformation

After ultraviolet irradiation of ETU on silica gel, Cruickshank & Jarrow (1973) found nine secondary reaction products. 2-Imidazolidinone was identified as the main degradation product and there were smaller amounts of 3-(2-imidazolin-2-yl)-2-imidazolidinethione (Jaffe's base). Other secondary reaction products of the photooxidation of ETU are 2-imidazoline and glycine (Ross & Crosby, 1973) via the intermediate hydantoin (IUPAC, 1977) (Fig. 4).

4.1 Soil

ETU degradation was found to be slower in autoclaved soils than in non-sterile soils (Kaufman & Fletcher, 1973), and only EU was identified. In biologically active soils, ETU was oxidized to carbon dioxide and four other degradation products, two of which were identified as hydantoin and Jaffe's base. Degradation of ETU to carbon dioxide in non-sterile soils was reported by Lyman & Lacoste (1974). These results indicate that ETU is oxidized under both biological and non-biological conditions to EU, which is considerably more stable than ETU and can be considered a major breakdown product. EU, however, can be oxidized photochemically, using a catalyst, to give glycine and carbon dioxide (Ross & Crosby, 1973), or microbially in soil. In this context, Jaffe's base might be considered as an intermediate product in ETU degradation.

According to Lyman & Lacoste (1974) and Rhodes (1977), half of the ETU (present at a concentration of 10 mg/kg) in Hagerstown silt loam soil was degraded to carbon dioxide in 22 days. Normal microbial carbon dioxide production was unaffected by ETU at this concentration. Because this value was determined on the basis of ^{14}C-carbon dioxide formation from ^{14}C-labelled precursor, it does not represent a half-life of ETU, since ^{14}C-carbon dioxide formation did not parallel the disappearance of labelled starting material from the soil. The actual half-life of ETU is less than one day.

According to Kaufman & Fletcher (1973), ETU is oxidized to EU, whereas carbon dioxide is only formed slowly. In Hagerstown silt loam, ETU at 2 or 20 mg/kg was entirely converted into EU

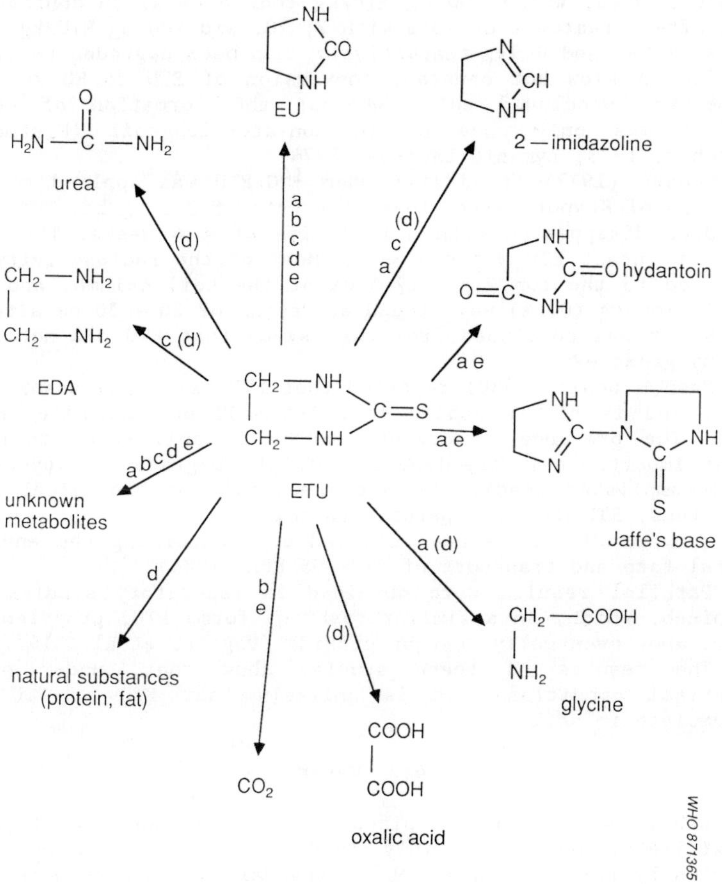

Fig. 4. Reaction products of ETU in biological and non-biological systems.
a = photodecomposition.
b = chemical oxidation.
c = plants.
d = animals.
e = soil.
Letters in parentheses () indicate proposed pathways.
From: IUPAC (1977).

within 2 days, while 200 mg ETU/kg took 8 days. In contrast, 4 days after treatment of soil with 2, 20, and 200 mg ETU/kg, only 43.4%, 8.9%, and 0.9%, respectively, had been degraded to carbon dioxide. A slow but constant conversion of ETU to EU was also found in autoclaved soil, whereas the formation of carbon dioxide was only observed in non-sterile soil (Kaufman & Fletcher, 1973; Lyman & Lacoste, 1974).

Rhodes (1977) found that when ^{14}C-ETU was applied to soil sections of Keyport silty loam at a rate of 2.2 kg/ha, total ^{14}C residues disappeared with a half-life of < 4 weeks. The half-life of intact ETU was < 1 week. Most of the radioactivity was confined to the top 2.5 - 12.5 cm of the soil column, and only small amounts (0.2%) were found at depths of 20 - 30 cm after 12 weeks. It was concluded from this study that ETU did not leach to any great extent.

Nash & Beall (1980) reported that ETU is weakly adsorbed to soil, and is highly mobile in moist soil but immobile in dry soil. The presence of organic matter in soil seems to be of great importance in the leaching of ETU. Degradation appears to be accomplished readily by both chemical and biological means and, thus, ETU does not persist in soil.

Many studies have been carried out concerning the environmental fate and transport of ETU (US EPA, 1984).

Parallel results were obtained in laboratory studies with propineb, which, in a similar fashion, forms PTU, propyleneurea (PU), and, eventually, carbon dioxide (Vogeler et al., 1977).

The results of these studies show that, under normal practical conditions, it is unlikely that ETU or PTU will accumulate in soil.

4.2 Water

ETU is stable in de-ionized water in the absence of photosensitizers, but is rapidly oxidized in their presence. In studies by Ross & Crosby (1973), several sensitizers were added at 10 mg/litre to a 25 mg/litre solution of ETU and exposed to sunlight. After 4 days with riboflavin as a sensitizer, the concentration of ETU was less than 5% of that in the control solution kept in darkness. To minimize microbial degradation, the procedure was repeated after filtering and boiling the water samples, with the same results. Furthermore, ETU degradation was investigated in several boiled samples of agricultural drainage water to which 0.5 mg ETU/litre had been added before irradiation. The results are given in Table 10.

Numerous samples of natural water were collected from rivers, lakes, and agricultural areas and, almost without exception, they were found to degrade ETU to EU in sunlight. The same samples degraded ETU in the dark but only after prior exposure to sunlight, indicating that stable photo-oxidants had

Table 10. Photodecomposition of ETU in agricultural waters[a]

Source	Irradiation	Remaining ETU (%)
Irrigation ditch (sugar beet)	3 days, lamp	10 - 20
	3 days, dark	100
Paddy flooding ditch (rice)	24 days, sun	25 - 50
	24 days, dark	100
Paddy (rice)	24 days, sun	10 - 25
	24 days, dark	100

[a] From: Ross & Crosby (1973).

been generated. The substances responsible for ETU oxidation were isolated and identified as the amino acids tryptophane and tyrosine. The pure amino acids also caused the conversion of ETU to EU in the light, apparently by their ability to form hydroperoxides or other strong oxidants (Ross & Crosby, 1973). As both the amino acids and photosensitizers such as acetone, riboflavin, and chlorophyll are known to occur world wide in water and soil, and this photolysis also has been shown to take place rapidly on a silica surface (Cruickshank & Jarrow, 1973), the degradation of ETU to harmless products in the field seems entirely plausible (IUPAC, 1977).

4.3 Plants

In studies in which the roots of corn, lettuce, tomato, and pepper seedlings were treated with ETU, it was rapidly absorbed by roots, translocated subsequently to the foliar tissues, and then degraded very rapidly; virtually no ETU was detectable after 20 days (Hoagland & Frear, 1976). When cucumber seedlings were exposed to aqueous solutions of nabam or suspensions of zineb or maneb, ETU was rapidly absorbed by the roots and translocated within the plants. ETU appeared to be stable for at least 2 weeks in seedlings and, inside the plant, a slow conversion of ETU into 2-imidazoline was detected (Vonk & Kaars Sijpesteijn, 1970, 1971).

In greenhouse studies, ^{14}C-ETU was applied either to the soil or to the leaves of 4-week-old potato plants and 8-week-old dwarf tomato plants. Radioactivity was monitored in various parts of the plant at different time intervals. The application of 40 mg ^{14}C-ETU/kg to the leaves of potato plants resulted in negligible radioactivity in the roots and tubers 60 - 90 days after application. The application of 17 - 22 kg/ha to soil around the base of the plants resulted in a negligible amount of radioactivity in the tubers, roots, and foliage of potato plants, 60 - 90 days later. Comparable results were obtained with tomato plants, using other dose levels and periods (Lyman & Lacoste, 1974).

After systemic uptake of ETU by plants, EU and 2-imidazoline were identified as metabolites. Surface deposits of ETU, which may have occurred as a result of EBDC treatment, formed an additional unidentified substance as the main metabolite and ethylene diamine (EDA). Propineb and PTU also formed an identical but unidentified major metabolite under similar conditions (Vogeler et al., 1977).

Nash (1975) reported the presence of 7 - 10 different degradation products in methanol extracts of soybeans after soil or foliar treatment with EBDCs, as well as after treatment with ETU. In these cases, EU was a degradation product.

More recently, Nash & Beall (1980) studied the fate of maneb and zineb in microagroecosystem chambers (Part A, section 6.3). ETU on the tomato fruit and leaves, and in the soil, water, and air was monitored for 100 days after treatment. ETU was detected at < 20 μg/kg on whole fruit after 3 days, but had completely disappeared after 3 weeks. The half-life of ETU was < 3 days.

5. ENVIRONMENTAL LEVELS AND HUMAN EXPOSURE

The amount of ETU in commercial formulations of EBDCs has been shown to increase with increasing temperature and humidity. ETU formation during storage appears to be greatest in maneb formulations (up to 14%), followed by zineb and mancozeb. The relative proportions of degradation products appear to be different for the various EBDCs.

Studies with propineb on apples and grapes, carried out by Vogeler et al. (1977), showed that PTU could be detected shortly after treatment, and that it was rapidly transformed to an unknown metabolite, together with small amounts of PU, and other unidentified reaction products.

5.1 Food and Drinking-Water

The residue levels of ETU are mainly below 0.1 mg/kg product following treatment (with different formulations) at the maximum recommended EBDC levels (Newsome, 1976; Phillips et al., 1977).

Studies in Canada and the USA have shown that vegetables, such as spinach, carrots, and potatoes, that have been treated with EBDCs contain high levels of ETU after cooking (Blazquez, 1973; Newsome & Laver, 1973; Watts et al., 1974). Snapbeans treated with maneb were found to contain ETU after commercial canning (US EPA, 1977). Farrow & Ralls (1970) demonstrated the disappearance of zineb, ziram, and maneb residues from spinach and apricots during normal canning operations. There have been many studies on ETU residues in crops, such as those of Sato & Tomizawa (1960) on zineb-treated cucumbers.

The levels of ETU formed from residues of mancozeb and polyram present on apples that were processed to make apple juice, apple sauce, and apple pomace were determined (IUPAC, 1977). The results showed that ETU residues were 0.17 mg/kg pomace and 0.05 mg/kg juice. A surprisingly high level of unchanged mancozeb remained in the pomace, despite heat treatment for 15 h at 150 °C. However, it seems that ETU residues diminish during storage.

Residues of intact ^{14}C-labelled ETU were found to diminish with time in canned tomato sauce, spinach, pickles, and apple sauce (Rose et al., 1980). EU and more polar products accounted for most of the ^{14}C-labelled residues. These polar materials were resistant to extraction and appeared to be bound.

A study sponsored by the US EPA (Phillips et al., 1977), on the effects of food processing on EBDC residues, confirmed and extended the results previously described. Washing the raw agricultural products prior to processing removed 33 - 87% of the EBDC residue and the majority of the ETU residue. An interesting result was that although almost instantaneous conversion of mancozeb to ETU took place in boiling water, field weathered residues of mancozeb appeared to be more resistant to degradation to ETU. A summary of the results for raw and processed material is given in Table 3.

5.2 Monitoring and Market-Basket Studies

A monitoring programme initiated in 1972 by the Canadian government showed that 33% of food samples contained detectable ETU residues. In particular, samples of canned spinach and orange peel had average values of 0.047 mg/kg product and 0.083 mg/kg, respectively (Pecka et al., 1975; US EPA, 1977).

Studies on the actual level of ETU in products prepared for commercial sale show it to be generally present in small amounts. The highest level, 0.61 mg/kg product, was found in canned peaches, while levels in orange peel, tomato paste, instant potatoes, strawberries, peaches, and cucumbers were less than 0.2 mg/kg product (US EPA, 1982a,b).

The US EPA has estimated an upper limit for dietary exposure to ETU in the general population of the USA to be 3.65 μg/kg body weight per day. This estimate is a maximum value, since it was assumed that residues are present at the tolerance level and that all of the EBDC residue is quantitatively converted to ETU. Using actual residue data and experimentally derived conversion factors, the US EPA estimated the dietary intake of ETU to be 0.24 μg/kg body weight per day.

In a market-basket study, over 500 samples of 34 foods were analysed, plus 26 samples of drinking-water. No water samples and only 21 of the food samples contained ETU residues (Gowers & Gordon, 1980). Exposure estimates based on market-basket studies range from 0.01 to 1 μg ETU/kg body weight per day (Gowers & Gordon, 1980; Rose et al., 1980).

Tomato products (203 samples) were analysed in another market-basket study, but none contained ETU (Gowers & Gordon, 1980).

A more realistic review of the actual exposure of the general population was obtained by a "table-top" study. Of 200 meals (some from homes and some from restaurants) which were analysed for ETU, none contained any residues (Gowers & Gordon, 1980).

6. KINETICS AND METABOLISM

6.1 Absorption, Distribution, and Excretion

ETU is rapidly absorbed from the gastrointestinal tract and cleared from the body in all the mammalian species that have been tested. After only 5 min, ETU appeared in the blood of rats administered an oral dose of 100 mg ^{14}C-ETU/kg body weight. Within 48 h, 82 - 99% of an oral dose was eliminated via the urine and about 3% via the faeces (Kato et al., 1976; Rose et al., 1980). Newsome (1974) and Ruddick et al. (1976a) found that approximately 70% was eliminated in the urine and 1% in the faeces. Comparable results were found for mice while, in monkeys, 55% was eliminated via the urine within 48 h, and less than 1.5% via the faeces (Allen et al., 1978).

To study the accumulation and elimination of radioactivity by the thyroid gland of rats dosed with ^{14}C-ETU, dose levels of 2 and 200 μg labelled ETU were administered daily for 14 days. In another study, rats were dosed with 0, 0.1, 1, 10, 50, or 100 mg ^{14}C-ETU/kg diet, daily, for 7 days. The first study showed that the concentration of ETU and/or its metabolites in the thyroid is dose dependent, and the second that the level of ^{14}C in the thyroid did not increase appreciably when the daily dose was increased above 50 mg/kg diet. Withdrawal of ETU from the diet led to an 80 - 94% reduction in the radioactivity in the thyroid after 17 days (Lyman & Lacoste, 1974).

ETU and its metabolites have been found to have a half-life of about 28 h in monkeys, 9 - 10 h in rats, and 5 h in mice (Rose et al., 1980).

In cows administered 1 mg ^{14}C-ETU/kg diet, Lyman (1971) found a small quantity of unchanged ETU in both the urine and the milk of the test animals. Higher levels of ^{14}C were detectable in metabolites, such as glycine and urea, and in the lactose and protein in the milk (Table 11).

6.2 Metabolic Transformation

It has been demonstrated that ETU degradation leads to traces of EU and other metabolites in the urine and that ^{14}C-carbon dioxide is exhaled following the administration of labelled ETU. Kato et al. (1976) suggested that the metabolites of ETU in the rat were produced primarily by fragmentation of the imidazoline ring and decarboxylation of the fourth and fifth carbon atoms. A small amount of radioactivity was also found in a protein fraction of rat fetal tissue. Ruddick et al. (1976a), however, concluded that ETU metabolism in the rat does not appear to result in any release of ^{14}C into the general

Table 11. ^{14}C Activity in the milk and urine of cows fed with 1 mg ^{14}C-ETU[a]/kg diet for 6 weeks

Substance	Milk Concentration (mg/litre)	Milk % of total ^{14}C	Urine Concentration (mg/litre)	Urine % of total ^{14}C
ETU	0.011	31	0.12	7
EU	0.0025	8	0.27	18
EDA	-	-	0.14	14
Glycine	-	-	-	6
Oxalic acid	-	-	-	12
Urea	-	-	-	11
Fat	-	3	-	-
Protein	-	18	-	-
Lactose	-	16	-	-
Total (%)		76		68

a From: Lyman (1971) and IUPAC (1977).

metabolic pool. Mice metabolize ETU to EU and other unknown metabolites, while cats metabolize it to *S*-methyl-ETU and EU.

Lyman (1971) detected EU, EDA, oxalic acid, glycine, and urea as major metabolites in cow urine. In addition, ^{14}C originating from ^{14}C-ETU was found in the protein and lactose in the milk (Table 11). From these results, it appears that the metabolism of ETU in ruminants is different from that in non-ruminants. The degradation products of ETU in plants are similar to those found in animals.

A summary of the secondary metabolites of ETU in biological and non-biological systems is given in Fig. 4 (Part B, section 4).

7. EFFECTS ON ORGANISMS IN THE ENVIRONMENT

Only limited information is available on the effects of ETU on organisms in the environment, and none is available concerning the impact of ETU on terrestrial organisms. Data concerning the toxicity for aquatic organisms of ETU and its breakdown product EU are summarized in Tables 4 and 5. From these results, ETU appears to have a low toxicity for bacteria, algae, crustacea, and fish. Because of the low partition coefficient (Table 6) and rapid biotransformation of ETU, bioaccumulation will be insignificant or absent (Van Leewen, 1986).

8. EFFECTS ON EXPERIMENTAL ANIMALS AND *IN VITRO* TEST SYSTEMS

8.1 Single Exposures

Lewerenz et al. (1975) reported an acute oral LD_{50} for the rat of 900 mg/kg body weight. The values determined by Graham & Hansen (1972) and Teramoto et al. (1978a) were 1832 mg/kg and 545 mg/kg body weight, respectively. Teramoto et al. (1978a) reported values of 3000 mg/kg body weight for the mouse and > 3000 mg/kg body weight for the hamster.

8.2 Short- and Long-term Exposures

Administration of ETU to laboratory rats causes enlargement of the thyroid gland. This effect was noted by Seifter & Ehrich (1948), and has since been confirmed in various short- and long-term studies. The accumulation of ETU in the thyroid gland is associated with biochemical and morphological effects comparable with those induced by known antithyroid drugs such as thiouracil.

Graham & Hansen (1972) fed rats with diets containing ETU at dose levels of 0, 50, 100, 500, or 750 mg/kg diet for 30, 60, 90, or 120 days, and at 100 mg/kg or more, thyroid changes were seen. Ross Hart & Valerio (1973) fed doses of up to 1000 mg ETU/kg diet to rats. An increase in thyroid weight was seen at 159 mg/kg or more, and the larger doses produced hyperplasia.

When Freudenthal et al. (1977) carried out studies on rats at dose levels of up to 625 mg ETU/kg diet for 30, 60, or 90 days, biochemical changes reflecting effects on thyroid function were observed. A dose of 125 mg/kg diet produced, after 30 days, a decrease in T_3, T_4, and serum PBI levels and an increase in TSH concentration. A dose of 25 mg/kg diet produced a decrease in T_4 content after 60 days. Thyroid weight was increased in all groups receiving 25 mg/kg or more. After 90 days, tumours (adenomas) were found in the 125 mg/kg group. The group receiving 625 mg/kg died within 7 weeks.

From these short-term studies, it can be concluded that the no-observed-adverse-effect level lies below 25 mg ETU/kg diet and is probably of the order of 5 mg/kg diet (equivalent to 0.25 mg/kg body weight).

8.3 Teratogenicity

ETU was administered orally to rats and rabbits in single daily doses of 0, 5, 10, 20, 40, or 80 mg/kg body weight. Rats were treated from 21 - 42 days before conception to day 15 of pregnancy, or on days 6 - 15 or 7 - 20 of pregnancy, whereas rabbits were treated on days 7 - 21 of pregnancy. ETU at dose levels of 10 mg or more induced meningoencephalocele, meningorrhagia, meningorrhea, hydrocephalus, obliterated neural canal, abnormal pelvic limb posture with equinovarus, and short, kinky tail in all rats. Fetal survival was not affected, and fetal growth was retarded only at 40 and 80 mg/kg. Rabbits showed an increased incidence of resorption sites and decreased brain weight at 80 mg/kg body weight, but no malformations (Khera, 1973).

ETU was studied in rats and mice for its ability to induce perinatal toxicity and in guinea-pigs and golden hamsters for its teratogenic potential. The ETU was administered by gavage and during organogenesis. Table 12 summarizes the prenatal treatments. Additional postnatal studies were performed on rats using extended treatment periods (ETU at 0, 20, 25, or 30 mg/kg body weight), including continuous exposure from day 7 of gestation through parturition to day 15 of lactation. The pups were weaned normally and postnatal studies on open-field behaviour were performed at 6 weeks. ETU at 80 mg/kg body weight induced maternal toxicity and reduced growth in the rat and was teratogenic for the rat, inducing substantial fetal effects at all dose levels above 10 mg/kg body weight. Gross defects were seen in the skeletal and the central nervous systems, and cleft palate was noted, mainly at the highest dose level. At 20 and 30 mg/kg, an increased incidence of hydrocephalus was the only defect noted. The maternal and fetal toxicity of ETU for the mouse, guinea-pig, and hamster was substantially less than for the rat. In mice, at the highest dose, an increase in maternal liver weight and in fetal supernumerary ribs was noted, but no effects were seen in other species (Chernoff et al., 1979; FAO/WHO, 1980b).

Table 12. Summary of prenatal treatments[a]

Compound	Species	Dose (mg/kg body weight)	Treatment (gestation days)
ETU	rat	0, 5, 10, 20, 30, 40, 80	7 - 21
	mouse	0, 100, 200	7 - 16
	hamster	0, 25, 50, 100	5 - 10
	guinea-pig	0, 50, 100	7 - 25

[a] Modified from: Chernoff et al. (1979).

ETU has been found to induce a variety of postnatal effects, including reduction or absence of maternal milk production, thereby causing pup mortality at doses of 30 mg/kg body weight or more. However, no significant dose-related behavioural abnormalities were observed in ETU-exposed pups (Chernoff et al., 1979; FAO/WHO, 1980b).

In studies by Khera & Tryphonas (1977), groups of pregnant rats were administered ETU at dose levels of 0, 15, 30, or 45 mg/kg body weight on day 15 of gestation and then subjected to a variety of test conditions to evaluate pre- and postnatal effects. Postnatal mortality occurred in pups from mothers treated with dose levels exceeding 15 mg/kg or pups cross-fostered to evaluate lactation exposure. All pups from mothers treated with 45 mg/kg died within 4 weeks of birth. A high incidence of hydrocephalus and microphthalmia was observed in pups of mothers treated with 30 mg/kg and these pups died within 6 weeks of birth. Motor defects observed in some survivors (16/65) of this group, were shown to result from the hydrocephalic condition, which was accompanied by atrophy of the cerebral cortex and subcortical white matter. These defects were found to be a direct result of *in utero* exposure to ETU and not of exposure during lactation (cross-fostered pups showed the same effects as pups weaned from treated dams). When mated with normal male rats, all female offspring of rats administered 30 mg/kg gave birth to normal offspring. The F_2 generation was not impaired, though some of the parents had neurological defects. In these studies, no effects on the parameters examined were observed at 15 mg/kg body weight (Khera & Tryphonas, 1977).

Teramoto et al. (1978a) investigated the teratogenicity of ETU in rats, mice, and hamsters. It was teratogenic when given orally to rats at 20 - 50 mg/kg body weight per day on days 6 - 15 of pregnancy and to hamsters at 270 - 810 mg/kg body weight per day on days 6 - 13 of pregnancy. However, no malformations were induced in mice up to a daily oral dose of 800 mg/kg body weight when given on days 7 - 15 of pregnancy. In hamsters, cleft palate, kinky tail, oligodactyly, and anal atresia were noted as gross external malformations. Skeletal examination revealed a high incidence of defects in the ribs and vertebral column, but no apparent defect was observed during visceral examination. An oral dose of 100 or 200 mg ETU/kg body weight given to pregnant rats consistently produced brain abnormalities in the fetuses, when given on day 12 or 13 of pregnancy, and forelimb abnormalities, when given on day 13 of pregnancy (Teramoto et al., 1978a). Histological studies revealed extensive cell necrosis in the brain and forelimbs of embryos 24 h after the treatment. These lesions were considered to be the main cause of the abnormalities observed. However, neither malformations nor cell necrosis were found in the fetuses that had been injected with 200 μg ETU/conceptus into the amniotic sac on day 12 of pregnancy (Teramoto et al., 1980). Studies with 2-^{14}C-ETU revealed that this dose was sufficient to test the direct effects of ETU on the embryos, since the incorporation of radioactive substance was five times higher in the embryos injected with 200 μg into the amniotic sac than it was in those embryos whose mothers were treated with an oral dose of 100 mg/kg body weight (Teramoto et al., 1980).

The teratogenic potential of the ETU metabolite 1-methylthiourea has been investigated by Teramoto et al. (1981). It caused almost the same types of malformations in rat fetuses when given orally to mothers at 250 - 500 mg/kg body weight on day 12 or day 14 of pregnancy as those observed following treatment with ETU. However, 1-methylthiourea did not induce malformations in mouse fetuses whose mothers were given an oral dose of 1000 mg/kg on day 10 of pregnancy. There is a structural similarity between 1-methylthiourea and ETU: C=S, and -NH- groups seem essential for producing teratogenic effects (Teramoto et al., 1981). However, the structure of ETU seems quite specific for the induction of teratogenicity since Ruddick et al. (1976b) tested 16 compounds related to ETU, including ethylenethiuram monosulfide, another metabolite, and only one, 4-methylenethiourea, was teratogenic.

8.4 Mutagenicity

Tests with a large number of *S. typhimurium* strains gave mostly negative results, though a few (weak) positive results were observed in the case of some strains of *S. typhimurium* (Shirasu et al., 1977). The addition of rat liver microsomes seemed to enhance the mutant reversion. Schüpbach & Hummler (1976, 1977) concluded that ETU appeared to induce base-pair mutations but not frameshift mutations in *S. typhimurium* TA 1530, although frameshift mutations appeared in *S. typhimurium* TA 98, TA 1537, and TA 1538 when exposed to ETU in the presence of dimethylsulfoxide (DMSO) and/or rat liver microsomes (Rose et al., 1980).

Teramoto et al. (1977) did not find mutagenicity with *S. typhimurium* TA 1536, TA 1537, TA 1538, G46, *E. coli* WP2 hcr$^+$ and hcr$^-$, or *B. subtilis* H17 rec$^+$ and rec$^-$ at concentrations of 10 000 μg ETU/plate. However, a weak reaction was seen with *S. typhimurium* TA 1535, and Seiler (1974) reported weak (dose-unrelated) mutagenicity in *S. typhimurium* strain G46.

ETU was also found to be mutagenic in a host-mediated assay of *S. typhimurium* TA 1530 when mice were dosed with 6000 mg ETU/kg body weight, but not at doses of 2000 mg/kg or less (Schüpbach & Hummler, 1977). Cytogenetic effects of ETU have been reported in bone marrow cells of mice and Chinese hamsters. On the other hand, there was no significant evidence to suggest that ETU was mutagenic in host-mediated assays of *S. typhimurium* G 46 or in tests with other strains of bacteria, rat bone marrow (including the micronucleus test) Chinese hamster DON cells, rat lymphocytes, or human fibroblast cells. Furthermore, ETU did not increase the frequency of dominant lethal mutations in rodents or *Drosophila melanogaster* (Seiler, 1973, 1974; Schüpbach & Hummler, 1977; Shirasu et al., 1977; Teramoto et al., 1978b; Rose et al., 1980). A large number of mutagenicity tests are summarized in a report of the US EPA (1984).

ETU has been tested in the hepatocyte DNA repair test, which is used to determine pro-carcinogenic potential as well as DNA damage. ETU did not induce DNA damage (Althaus et al., 1982) nor cause chromosomal damage in cultured rat liver cells, and it did not induce chromatid exchange in CHO cells *in vitro* or in mice *in vivo*. A micronucleus test with mice bone marrow cells *in vivo* also gave negative results (De Serres & Ashby, 1981).

This evidence indicates that ETU is generally not mutagenic, especially in mammalian test systems.

8.5 Carcinogenicity

The carcinogenicity of ETU has been evaluated by IARC (1974, 1982). It was classified in group 2B, i.e., limited evidence for activity in short-term tests; sufficient evidence for carcinogenicity in animals; inadequate evidence for carcinogenicity in human beings.

ETU has been studied for oncogenic potential in mice, rats, and hamsters.

8.5.1 Mouse

In a comprehensive programme screening chemicals for carcinogenicity, two strains of hybrid mice (X and Y) were given 215 mg ETU/kg body weight from day 7 until weaning, and thereafter 646 mg/kg diet for more than 18 months. In the X strain [(C_{57}Bl/6XC_3H/Anf)F_1], the incidence of lung tumours in the ETU-treated females was higher than that of the control group (3/18 versus 3/87), but it was lower than that of the controls (0/16 versus 1/90) in the Y strain [(C_{57}Bl/6XAKR)F_1]. In the males, the incidence was higher in the ETU-treated animals (3/18 versus 1/90). The incidence of lymphomas was slightly increased in treated Y-strain females. The hepatoma incidence in the ETU-treated groups of both strains was significantly higher than that of the control group (in the X strain, 14/18 and 18/18, for males and females respectively; in the Y strain, 18/18 and 9/18; in controls, 0/18 and 3/18). The thyroid glands were not examined for histopathological changes (Innes et al., 1969).

Graham et al. (1975), found that ETU induced thyroid hyperplasia and other research groups have confirmed this finding.

8.5.2 Rat

Ulland et al. (1972) and Weisburger et al. (1981) fed groups of 26 male and female Charles River-CD rats diets containing 0, 175, or 350 mg ETU (97%)/kg diet. Five females and five males of the high-dose group were killed after 18 months and the remainder after 24 months. Hyperplastic goitre, solid cell adenomas, and thyroid (follicular or papillary) carcinomas were found. Two of the animals also had lung tumours, which might have been metastases. The thyroid tumour incidence was dose dependant; (in the 175 mg/kg group, it was 3/26 and 3/26, for males and females, respectively, and in the 350 mg/kg group it was 17/26 and 8/26). No thyroid carcinomas were observed in the control animals. A few of the treated rats had hyperplastic nodules in the liver.

Graham et al. (1973, 1975) studied the long-term effects on the thyroid gland of ETU ingestion. Five groups of 68 male and 68 female Charles River rats were fed ETU at levels of 0, 5, 25, 125, 250, or 500 mg/kg diet for 2 years. Growth depression was evident at the highest dose level. The thyroid/body weight ratio was significantly increased at 250 and 500 mg/kg, and slightly increased at 125 mg/kg after 24 months. Thyroidal uptake of ^{131}iodine per mg tissue was significantly decreased in male rats fed 500 mg ETU/kg diet for 18 or 24 months. The thyroids of females fed at the three highest dose levels were hypofunctioning at 6 months, and hyperfunctioning at 12 months, and at 24 months thyroid function was similar to that of the controls. At the two highest dose levels (250 and 500 mg/kg), thyroid adenomas and carcinomas were induced. At all lower dose levels hyperplasia occurred more frequently than in the controls, but there were no adenomas or carcinomas. No increase in liver tumours was observed in this study.

Gak et al. (1976) studied the effects of feeding rats with 0, 5, 17, 60, or 200 mg ETU/kg diet for 24 months. Body weight, food consumption, serum enzyme activities (e.g. glutamic pyruvic transaminase, alkaline phosphatase), hepatic enzyme activities (glutamic pyruvic transaminase, alkaline phosphatase, glucose-6-phosphate dehydrogenase), cholesterol levels, weights of thyroid and other organs, and histopathology were studied. Hypercholesterolemia was found at dose levels of 5 mg/kg and above. At 60 mg/kg or more, a significant increase in thyroid tumours was found, but at lower levels the tumour incidence was not significantly different from that of the controls.

8.5.3 Hamster

Gak et al. (1976) studied the effect of 0, 5, 17, 60, or 200 mg ETU/kg diet on hamsters for 18 months. Growth, food intake, biochemical parameters in the serum and liver, organ weights, and histology were studied. A significant increase in thyroid tumours was found at 60 mg ETU/kg or more, but at lower doses values were not significantly different from those of the control group.

8.6 ETU in Combination with Nitrite

When the mutagenicity of ETU was assayed before and after nitrosation with sodium nitrite under acid conditions, nitrosation was found to cause a 160-fold increase in the number of revertant colonies of *S. typhimurium TA 1535* (Shirasu et al., 1977). The interactive mutagenicity of ETU and nitrite was also found in the mouse dominant lethal test by Teramoto et al. (1978b). However, no dominant-lethal mutations were induced in a group of mice treated with 30 mg ETU plus 10 mg nitrite/kg body weight. A large increase in pre-implantation losses was noted 5 and 6 weeks after completing a 5-day treatment of males with a combined oral dose of 150 mg ETU/kg and 50 mg sodium nitrite/kg body weight.

8.7 Mechanisms of Toxicity; Mode of Action

The biochemical changes induced by antithyroid drugs include reduced production of thyroid hormones (T_3 and T_4), followed by increased production of TSH in response to low thyroid hormone levels in the blood. Pathological changes in the thyroid gland begin with diffuse microfollicular hyperplasia, and are followed by diffuse and nodular hyperplasia and later by nodular hyperplasia with papillary and cystic changes induced by the TSH. If hyperstimulation of the thyroid by TSH is severe and prolonged, it provides conditions conducive to the formation of tumours.

Antithyroid drugs can inhibit T_4 production in various ways. The chemical similarity of ETU to thiourea and thiouracil suggests that ETU acts by blocking the iodination of thyroxine precursors, thus reducing the synthesis of the thyroid hormones. Iodide peroxidase catalyses the iodination of tyrosine and the coupling of the resultant iodotyrosyl residues to produce the active hormones T_3 and T_4.

Graham & Hansen (1972) found that ETU inhibited iodide peroxidase *in vitro*. The resulting decreased level of thyroid hormones causes stimulatory feedback of the pituitary gland and consequently an increased release of TSH (Rose et al., 1980).

Lu & Staples (1978) studied the influence of ETU in pregnant hypothyroid and euthyroid rats to determine whether ETU teratogenicity occurs as a result of altered maternal thyroid function. Doses of 40 mg ETU/kg body weight, administered on days 7 - 15 of gestation, resulted in 84 - 100% of the fetuses in all treated groups being malformed, regardless of the thyroid status of the dams. The authors concluded that the thyroid status of the mother is not of importance in causing teratogenic effects.

Rose et al. (1980) reported that the effects of feeding rats 125 - 625 mg ETU/kg diet for 2 - 12 weeks, which included thyroid hyperplasia and dose related suppression of serum T_3 and T_4 (with corresponding TSH elevation), were reversible within 22 weeks of placing on control diets.

Long-term studies using ETU showed significantly increased thyroid/body weight ratios in rats fed 125, 250, or 500 mg/kg diet for periods of up to 2 years (Graham et al., 1975). This effect was not reversed in rats placed on a control diet after 66 weeks of continuous exposure to 5 - 500 mg ETU/kg diet. It is likely that by that time the thyroid was severely damaged.

In studies by Arnold et al. (1982, 1983), decreased levels of serum thyroid hormones and increased thyroid weights were reversed in Sprague Dawley rats fed diets containing 0, 75, 100, or 150 mg ETU/kg diet for 7 weeks. The reversibility of microscopic changes in the thyroids of male rats exposed to ETU was studied. The rats were fed diets containing 75 or 150 mg ETU/kg diet for 7 - 82 weeks and then returned to a control diet for periods ranging from 2 to 42 weeks. The severity and extent of reversibility of thyroid hyperplasia were found to depend on the duration of exposure to ETU. Above a certain threshold, hyperplasia did not regress significantly.

Numerous studies with ETU suggest that the rat is more sensitive than other species to the effects of the thyroid. A recent study with propylthiouracil, a thyroid inhibitor with a mode of action similar to that of ETU, has confirmed that monkeys are much less sensitive than rats. The sensitivity difference was not quantified *in vivo*, but, in an *in vitro* study, the concentration of inhibitor required to produce the same level of thyroid peroxidase inhibition was approximately 100 times greater for monkey enzyme than it was for rat enzyme (Takayama et al., 1986).

8.8 Propineb and Propylenethiourea (PTU)

8.8.1 General

The toxicology of propineb was reviewed at JMPR meetings in 1977, 1980, and 1983. Because of concern expressed at the 1977 meeting regarding the potential for thyrotoxicity and tumourigenicity of propylenethiourea (PTU), a breakdown product of propineb, the meeting estimated only a temporary ADI for man. Further evaluation of propineb was postponed pending the submission of additional data. Data submitted for evaluation in 1985 consisted of long-term mouse and rat studies, mutagenicity studies, and a special study into the effects of PTU on DNA. In addition, data previously submitted for evaluation in 1983 were re-examined. These data included several studies on propineb (acute toxicity studies, a short-term study on thyroid function in rats, mutagenicity studies, and an oncogenicity study on mice) and on PTU (pharmacokinetic studies on rats and a long-term thyroid function study on rats) (FAO/WHO, 1986a,b).

8.8.2 Toxicological information

An oncogenicity study on mice with propineb indicated increased hepatocellular adenomas in male mice and increased pulmonary adenomas in female mice at 800 mg/kg diet, the highest dose level tested. Thyroid tumours were not induced in treated mice in this study. A no-observed-adverse-effect level for non-neoplastic effects could not be determined in this study, owing to insufficient data (FAO/WHO, 1986a, b).

In a long-term study into the effects of PTU on mice, an increased incidence in male mice of hepatocellular adenomas was observed at 1000 mg/kg diet (the highest dose level tested) and of hepatocellular carcinomas at 10 mg/kg diet or more. In the same study, increased incidences of hepatocellular adenomas and carcinomas were observed in female mice at 100 mg/kg diet or more. Thyroid tumours attributable to PTU were not observed, but increased thyroid hypercellularity was noted in male mice at 1000 mg/kg diet (FAO/WHO, 1986a, b).

In long-term rat studies with propineb, previously reviewed by the JMPR, an increased incidence of benign thyroid tumours was observed at 1000 mg/kg diet or more. Non-neoplastic thyroid effects were observed in the same study at 100 mg/kg diet or more. In another study, increased liver and kidney weights were observed at 100 mg/kg diet or more and a no-observed-adverse-effect level of 10 mg/kg diet was determined. In a long-term study on the effects of PTU on rats, thyroid tumours attributable to PTU were only found at 1000 mg/kg diet. Goitrogenic effects in the thyroid were observed, however, at dose levels as low as 1 mg/kg diet, the lowest dose level tested. A no-observed-adverse-effect level could not be determined in this study (FAO/WHO, 1986a, b).

Short-term studies on the effects of propineb on thyroid function in rats did not establish an unequivocal no-observed-adverse-effect level for effects on the thyroid. In a long-term study with PTU, effects on thyroid function were observed at 1000 mg/kg diet, but at lower dose levels effects were ambiguous. Pharmacokinetic studies on rats demonstrated preferential uptake of radioactivity from ^{14}C-labelled PTU by the thyroid (FAO/WHO, 1986a, b).

Mutagenicity studies on propineb and PTU produced negative or inconclusive results. However, PTU has been shown to increase DNA synthesis in mouse spleen cells, but it did not bind to mouse liver cell DNA.

In view of the carcinogenic response to PTU in the liver of mice and the lack of a no-observed-adverse-effect level for the effects of propineb on the thyroid in a long-term study on mice or short-term studies on rats, or for PTU in a long-term study on rats, the JMPR recommended that the temporary ADI for propineb should be withdrawn.

In view of the established carcinogenic potential of this compound, the meeting recommended that propineb should not be used where its residues can arise in food (FAO/WHO, 1986a,b).

9. EFFECTS ON MAN

9.1 Epidemiological Studies

Smith (1976) conducted a detailed study involving 1929 workers in rubber-compounding plants in Birmingham, England. No thyroid cancers were found in the health records of these workers.

Clinical examinations and thyroid function tests were carried out over a period of 3 years on eight process workers and five mixers in a factory producing ETU in the United Kingdom. Matched controls were also examined. The results showed that the exposed mixers, but not the process workers, had significantly lower levels of T_4 in their blood compared with the controls. No effect was found on TSH or thyroid-binding globulin (Smith, 1984).

PREVIOUS EVALUATIONS BY INTERNATIONAL BODIES

The Joint FAO/WHO Meeting on Pesticide Residues (JMPR) and the International Agency for Research on Cancer (IARC) have evaluated the toxicity and carcinogenicity data for various dithiocarbamates on several occasions. Annex III includes an overview of the JMPR meetings in which these compounds, ETU, and PTU have been evaluated, with their references, together with the WHO recommended classification of pesticides by hazard for individual dithiocarbamates. The existence of IARC evaluations and the availability of WHO/FAO Data Sheets and IRPTC Data Profiles and Legal Files are also indicated. These documents include more detail concerning the product and legal aspects, toxicological evaluation, and residues of individual dithiocarbamates in different food items.

REFERENCES

AASETH, J., ALEXANDER, J., & WANNAG, A. (1981) Effect of thiocarbamate derivatives on copper, zinc, and mercury distribution in rats and mice. *Arch. Toxicol.*, **48**: 29-39.

ALDRIDGE, W.N. & MAGOS, L. (1978) *Carbamates, thiocarbamates, dithiocarbamates*, Luxembourg, Commission of the European Communities (Report No. V/F/1/78/75 EN).

ALLEN, J.R., VAN MILLER, J.P., & SEYMOUR, J.L. (1978) Absorption, tissue distribution, and excretion of ^{14}C-ETU in the rhesus monkey and the rat. *Chem. Pathol. Pharmacol.*, **20**: 109-115.

ALTHAUS, F.R., LAWRENCE, S.D., SATTLER, G.L., LONGFELLOW, D.G., & PITOT, H.C. (1982) Chemical quantification of unscheduled DNA synthesis in cultured hepatocytes on an assay for the rapid screening of potential chemical carcinogens. *Cancer Res.*, **42**: 3010-3015.

ALVAREZ, M., CREEKMORE, R.W., & ROSEN, R.T. (1973) Structure of ethylenethiuram monosulfide. *Tetrahedron Lett.*, **12**: 939-942.

ANTONOVICH, E.A., CHERNOV, O.V., SAMOSH, L.V., MARTSON, L.V., PILINSKAYA, M.A., KURINNY, L.I., VEKSHTEIN, M.S., MARTSON, V.S., BALIN, P.N., & KHITSENKO, I.I. (1972) [A comparative toxicological assessment of dithiocarbamates.] *Gig. i. Sanit.*, **37**: 25-30 (in Russian).

ARNOLD, D.L., BICKIS, M.G., NERA, E.A., MCGUIRE, P.F., & MUNRO, I.C. (1982) Assessing the reversibility of thyroid lesions that result from feeding ETU. *Toxicologist*, **2**(1): 90 (Abstract of Presentation at the SOT Meeting 1982).

ARNOLD, D.L., KREWSKI, D.R., JUNKINS, D.B., MCGUIRE, P.F., MOODIE, C.A., & MUNRO, I.C. (1983) Reversibility of ethylenethiourea-induced thyroid lesions. *Toxicol. appl. Pharmacol.*, **67**: 264-273.

ASHWORTH, R., DE B., HENRIET, J., LOVETT, J.F., & MARTIJN, A., ed. (1980) *CIPAC handbook - Volume 1A: Analysis of technical and formulated pesticides*, Cambridge, United Kingdom, W. Heffer and Sons Limited. England.

AYANABA, A., VERSTRAETE, W., & ALEXANDER, M. (1973) Formation of dimethylnitrosamine, a carcinogen and mutagen, in soils treated with nitrogen compounds. *Soil Sci. Soc. Am. Proc.*, **37**: 565-568.

BABINI, G. (1966) [A case of erythroderma caused by zinc dithiocarbamate.] *Arch. Ital. Dermatol. Venereol. Sessuol.*, **34**: 230/238 (in Italian).

BALIN, P.N. (1970) [Experimental data on the blastomogenic activity of the fungicide maneb.] *Vrach. Delo*, **4**: 21-24 (in Russian).

BEN-DYKE, R., SANDERSON, D.M., & NOAKES, D.N. (1970) Acute toxicity data for pesticides. *World Rev. Pest. Control*, **9**: 119-127.

BENES, V. & SRAM, R. (1969) Mutagenic activity of some pesticides in *Drosophila melanogaster*. *Ind. Med. Surg.*, **38**: 50-52.

BENSON, W.R., ROSS, R.D., CHEN JO-YUN, R., BARRON, R.P., & MASTBROOK, D. (1972) Structure of ethylenethiuram monosulfide. *J. Assoc. Off. Anal. Chem.*, 55(1): 44-46.

BLACKWELL-SMITH, R., Jr, FINNEGAN, J.K., LARSON, P.S., SAHYOUN, P.F., DREYFUSS, M.L., & HAAG, H.B. (1953) Toxicologic studies on zinc and disodium ethylenebisdithiocarbamates. *J. Pharmacol. exp. Therap.*, **109**: 159-166.

BLAZQUEZ, C.H. (1973) Residue determination of ethylenethiourea (2-imidazol-idinethione) from tomato foliage, soil, and water. *J. agric. food Chem.*, **21**: 330-332.

BOGARTYKH, T.A., ORLOV, N.V., & KUZNETSOVA, T.V. (1979) Activity of lactate dehydrogenase and glucose-6-phosphate dehydrogenase in tissue of the testes under the action of some pesticides. *Vopr. Pitan.*, **5**: 49-53.

BONTOYAN, W.R. & LOOKER, J.G. (1973) Degradation of commercial ethylenebisdithiocarbamate formulations to ethylenethiourea under elevated temperature and humidity. *J. agric. food Chem.*, **21**: 338-342.

BONTOYAN, W.R., LOOKER, J.B., KAISER, T.E., GIANG, P., & OLIVE, B.M. (1972) Survey of ethylenethiourea in commercial ethylenebisdithiocarbamate formulations. *J. Assoc. Off. Anal. Chem.*, **55**: 923-925.

BOTTOMLEY, P., HOODLESS, R.A., & SMART, N.A. (1985) Review of methods for the determination of ethylenethiourea (imidazolidine-2-thione) residues. *Residue Rev.*, **95**: 45-89.

CAVANAGH, J.B. (1973) Peripheral neuropathy caused by chemical agents. *Crit. Rev. Toxicol.*, **2**: 365-417.

CHARKES, N.D., BRAVERMAN, L.E., PENKO, K.F., GOWERS, D.S., GORDON, C.F., LIPWORTH, L., & MALMUD, L.S. (1985) Thyroid function in male workers manufacturing dithane, an agricultural fungicide, and in men not exposed to dithane. In: *Proceedings of the Ninth International Thyroid Congress, Sao Paulo, Brazil, 1-6 September 1985*, p. 89.

CHERNOFF, N., KAVLOCK, R.J., ROGERS, E.H., CARVER, B.D., & MURRAY, S. (1979) Perinatal toxicity of maneb, ethylene-thiourea, and ethylene bisisothiocyanate sulfide in rodents. *J. Toxicol. environ. Health*, **5**: 821-834.

CHERNOV, O.V. & KHISTENKO, I.I. (1969) Blastomogenic properties of some derivatives of dithiocarbamic acid. *Vopr. Onkol.*, **15**(4): 71-74.

CHERNOV, O.V. & KHISTENKO, I.I. (1973) Some data on the distribution and elimination of ziram-^{35}S from the organism of warm-blooded animals. In: *Hygiene of using toxicology of pesticides and the clinical picture of intoxication*, Moscow, Medizina, **Vol. 10**, pp. 250-254. (in Russian).

CLARKE, D.G., BAUM, H., STANLEY, E.L., & HESTER, W.F. (1951) Determination of dithiocarbamates. *Anal. Chem.*, **23**(12): 1842-1846.

CONKIN, R.A. & GLEASON, L.S. (1964) Vegadex. In: Zweig, G., ed. *Analytical methods for pesticides, plant growth regulation and food additives*, New York, London, Academic Press, **Vol. 4**, pp. 249-257.

CRUICKSHANK, P.A. & JARROW, H.C. (1973) Ethylenethiourea degradation. *J. agric. food Chem.*, **21**(3): 333-335.

DALVI, R.R., POORE, R.E., & NEAL, R.A. (1974) Studies of the metabolism of carbon disulfide by rat liver microsomes. *Life Sci.*, **14**: 1785-1796.

DANIELSSON, B.R.G., OSKARSSON, A., & DENCKER, L. (1984) Placental transfer and fetal distribution of lead in mice after treatment with dithiocarbamates. *Arch. Toxicol.*, **55**: 27-33.

DAVYDOVA, T.B. (1973) [The influence of tetramethylthiuram disulfide on the estrous cycle and genital function of animals when inhaled.] *Gig. i Sanit.*, **38**(4): 108-110 (in Russian).

DEKHUYZEN, H.M., VONK, J.W., & KAARS SIJPESTEIJN (1971) Terminal residues of dithiocarbamate fungicides. In: Tahori, A.S., ed. *Proceedings of the International Symposium on Pesticide Terminal Residues, Tel Aviv, Israel,* 1971, London, Butterworth, pp. 233-241.

DE MATTEIS, F. & SEAWRIGHT, A.A. (1973) Oxidative metabolism of carbon disulfide by the rat. Effect of treatments which modify the liver toxicity of carbon disulfide. *Chem.-biol. Interact.,* **7**: 375-388.

DE SERRES, F.S. & ASHBY, J.C. (1981) *Report of the International Collaborative Program for Progress in Mutation Research,* **Vol. 1**, Amsterdam, Oxford, New York, Elsevier North-Holland Inc.

DISHOVSKI, H. & IVANOVA-CHEMISHANSKA, L. (1979) Ultrastructural changes in the neocortex of rats after chronic dithiocarbamate Antracol. *Activ. nerv. sup. (Praha),* **21**(4): 279.

EDINGTON, N. & HOWELL, J.McC. (1966) Changes in the nervous system of rabbits following the administration of sodium diethyldithiocarbamate. *Nature (Lond.),* **210**(5040): 1060-1062.

EDINGTON, N. & HOWELL, J.McC. (1969) The neurotoxicity of sodium diethyldithiocarbamate in the rabbit. *Acta neuropathol. (Berlin),* **12**: 339-347.

EISENBRAND, G., UNGERER, O., & PREUSSMANN, R. (1974) Rapid formation of carcinogenic *N*-nitrosamines by interaction of nitrite with fungicides derived from dithiocarbamic acid *in vitro* under simulated gastric conditions and *in vivo* in the rat stomach. *Food Cosmet. Toxicol.,* **12**: 229-232.

ELDJARN, L. (1950) The metabolism of tetraethylthiuram disulfide (Antabuse, Aversan) in the rat, investigated by means of radioactive sulfur. *J. clin. lab. Invest.,* **2**: 198-201.

ELESPURU, R.K. & LIJINSKY, W. (1973) The formation of carcinogenic nitroso compounds from nitrite and some types of agricultural chemicals. *Food Cosmet. Toxicol.,* **11**: 807-817.

ENGST, R. & SCHNAAK, W. (1967) [Studies on the metabolism of the ethylenebisdithiocarbamate fungicides maneb and zineb.] *Z. Lebensm. Unters. Forsch.,* **134**(4): 216-221 (in German).

ENGST, R. & SCHNAAK, W. (1969a) [Polarographic determination of ethylene bisthiuram monosulfide.] *Z. anal. Chem.,* **248**: 321-324 (in German).

ENGST, R. & SCHNAAK, W. (1969b) [Analysis of ethylene bisthiuram monosulfide.] *Z. anal. Chem.*, **244**: 254-256 (in German).

ENGST, R. & SCHNAAK, W. (1969c) [Studies on the metabolism of the ethylenebisdiothiocarbamate fungicides maneb and zineb (polarographic determination of the undecomposed fraction of active ingredient).] *Z. Lebensm. Unters. Forsch.*, **139**(3): 149-152 (in German).

ENGST, R. & SCHNAAK, W. (1970a) [Studies on the metabolism of the ethylenebisdiothiocarbamate fungicides maneb and zineb.] *Z. Lebensm. Unters. Forsch.*, **143**: 99-103 (in German).

ENGST, R. & SCHNAAK, W. (1970b) [Studies on the metabolism of the ethylenebisdiothiocarbamate fungicides maneb and zineb.] *Z. Lebensm. Unters. Forsch.*, **144**(2): 81-85 (in German).

ENGST, R. & SCHNAAK, W. (1974) Residues of dithiocarbamate fungicides and their metabolites on plant foods. *Residue Rev.*, **52**: 45-67.

ENGST, R., SCHNAAK, W., & LEWERENZ, H.J. (1971) [Studies on the metabolism of the fungicidal ethylenebisdiothiocarbamatesmaneb, zineb and nabam. V. The toxicology of breakdown products.] *Z. Lebensm Untersuch.*, **146**, 91-97.

FAHRIG, R. (1974) Comparative mutagenicity studies with pesticides. In: Montesano, R. & Tomatis, L., ed. *Chemical carcinogenesis assays*, Lyons, International Agency for Research on Cancer, pp. 161-181 (IARC Scientific Publication No. 10).

FAO (1985) *Production yearbook 1984*, Rome, Food and Agriculture Organization of the United Nations.

FAO/WHO (1964) *Evaluation of the toxicity of pesticide residues in food. Report of a Joint Meeting of the FAO Committee on Pesticides in Agriculture and the WHO Expert Committee on Pesticide Residues*, Geneva, World Health Organization (FAO Meeting Report No. PL:1963/13; WHO/Food Add./23).

FAO/WHO (1965a) *Evaluation of the toxicity of pesticide residues in food. Report of the Second Joint Meeting of the FAO Committee on Pesticides in Agriculture and the WHO Expert Committee on Pesticide Residues*, Geneva, World Health Organization (FAO Meeting Report No. PL:1965/10; WHO Food Add./26.65).

FAO/WHO (1965b) *Evaluation of the toxicity of pesticide residues in food*, Geneva, World Health Organization (FAO Meeting Report No. PL:1965/10/1; WHO Food Add./27.65).

FAO/WHO (1968a) *Pesticide residues in food. Report of the 1967 Joint Meeting of the FAO Working Party of Experts on Pesticide Residues and the WHO Expert Committee on Pesticide Residues*, Geneva, World Health Organization (FAO Meeting Report No. PL;1967/M/11; WHO Technical Report Series No. 391).

FAO/WHO (1968b) *1967 Evaluation of some pesticide residues in food*, Geneva, World Health Organization (FAO PL: 1967/M/1/1; WHO Food Add./68.30).

FAO/WHO (1971a) *Pesticide residues in food. Report of the 1970 Joint Meeting of the FAO Working Party of Experts on Pesticide Residues and the WHO Expert Committee on Pesticide Residues*, Geneva, World Health Organization (FAO Agricultural Studies No. 87; WHO Technical Report Series No. 474).

FAO/WHO (1971b) *1970 Evaluations of some pesticide residues in food*, Geneva, World Health Organization (AGP 1979/M/12/1; WHO Food Add./71.42).

FAO/WHO (1975a) *Pesticide residues in food. Report of the 1974 Joint Meeting of the FAO Working Party of Experts on Pesticide Residues and the WHO Expert Committee on Pesticide Residues*, Geneva, World Health Organization (FAO Agricultural Studies No. 97; WHO Technical Report Studies No. 574).

FAO/WHO (1975b) *1974 Evaluations of some pesticide residues in food*, Geneva, World Health Organization (AGP 1974/M/11; WHO Pesticide Residues Series No. 4).

FAO/WHO (1978a) *Pesticide residues in food. Report of the 1977 Joint Meeting of the FAO Panel of Experts on Pesticide Residues and the Environment and the WHO Expert Group on Pesticide Residues*, Rome, Food and Agriculture Organization of the United Nations (FAO Plant Production and Protection Paper No. 10 Rev.).

FAO/WHO (1978b) *1977 Evaluations of some pesticide residues in food*, Rome, Food and Agriculture Organization of the United Nations(FAO Plant Production and Protection Paper No. 10 Sup.).

FAO/WHO (1980a) *Pesticide residues in food. Report of the 1979 Joint Meeting of the FAO Panel of Experts on Pesticide Residues in Food and the Environment and the WHO Expert Group on Pesticide Residues*, Rome, Food and Agriculture Organization (FAO Plant Production and Protection Paper No. 20).

FAO/WHO (1980b) *1979 Evaluations of some pesticide residues in food*, Rome, Food and Agriculture Organization of the United Nations (FAO Plant Production and Protection Paper No. 20 Sup.).

FAO/WHO (1981a) *Pesticide residues in food. Report of the 1980 Joint Meeting of the FAO Panel of Experts on Pesticide Residues in Food and the Environment and the WHO Expert Group on Pesticide Residues*, Rome, Food and Agriculture Organization of the United Nations (FAO Plant Production and Protection Paper No. 26).

FAO/WHO (1981b) *1980 Evaluations of some pesticide residues in food*, Rome, Food and Agriculture Organization of the United Nations (FAO Plant Production and Protection Paper No. 26 Sup.).

FAO/WHO (1984a) *Pesticide residues in food. Report of the 1983 Joint Meeting of the FAO Panel of Experts on Pesticide Residues in Food and the Environment and the WHO Expert Group on Pesticide Residues*, Rome, Food and Agriculture Organization of the United Nations (FAO Plant Production and Protection Paper No. 56).

FAO/WHO (1984b) *1983 Evaluations of some pesticide residues in food*, Rome, Food and Agriculture Organization of the United Nations (FAO Plant Production and Protection Paper No. 61).

FAO/WHO (1985a) *Pesticide residues in food. Report of the 1984 Joint Meeting of the FAO Panel of Experts on Pesticide Residues in Food and the Environment and the WHO Expert Group on Pesticide Residues*, Rome, Food and Agriculture Organization of the United Nations (FAO Plant Production and Protection Paper No. 62).

FAO/WHO (1985b) *1984 Evaluations of some pesticide residues in food*, Rome, Food and Agriculture Organization of the United Nations (FAO Plant Production and Protection Paper No. 67).

FAO/WHO (1986a) *Pesticide residues in food. Report of the 1985 Joint Meeting of the FAO Panel of Experts on Pesticide Residues in Food and the Environment and the WHO Expert Group on Pesticide Residues*, Rome, Food and Agriculture Organization of the United Nations (FAO Plant Production and Protection Paper No. 68).

FAO/WHO (1986b) *1985 Evaluations of some pesticide residues in food, Part II - Toxicology*, Rome, Food and Agriculture Organization of the United Nations (FAO Plant Production and Protection Paper No. 72).

FARROW, R.P. & RALLS, J.W. (1970) *Investigation of changes in pesticide residues during processing and storage of fruit and vegetables*, National Canners Association, 109 pp (Report No. 5).

FISHBEIN, L. (1975) *Chromatography of environmental hazards. III. Pesticides*, Amsterdam, Oxford, New York, Elsevier Science Publishers.

FITZHUGH, O.G., WINTER, W.J., & NELSON, A.A. (1952) Some observations on the chronic toxicity of antabuse (tetraethylthiuramdisulfide). *Fed. Proc.*, **11**: 345-346.

FREUDENTHAL, R.I., KERCHNER, G.A., PERSING, R.L., BAUMEL, J., & BARON, R.L. (1977) Subacute toxicity of ethylene bisisocyanate sulfide in the laboratory rat. *J. Toxicol. environ. Health*, **2**: 1067-1077.

GAK, J.C., GRAILLOT, C., & TRUHAUT, R. (1976) Différence de sensibilité du hamster et du rat vis-à-vis des effets de l'administration à long terme de l'éthylène thiourée. *Eur. J. Toxicol.*, **9**: 303-312.

GARDNER-THORPE, C. & BENJAMIN, S. (1971) Peripheral neuropathy after disulfiram administration. *J. Neurol. Neurosurg. Psychiatr.*, **34**: 253-259.

GESSNER, T. & JAKUBOWSKI, M. (1972) Diethyldithiocarbamic acid methyl ester, a metabolite of disulfiram. *Biochem. Pharmacol.*, **21**: 219-230.

GOWERS, D.S. & GORDON, C.F. (1980) *Some public health aspects of the manufacture and use of zinc and manganese ethylenebisdithiocarbamate fungicides* Poznan, Poland, Institut Ochrony Roslin, February 1979 pp. 491-522.

GRAHAM, S.L. & HANSEN, W.H. (1972) Effects of short-term administration of ethylenethiourea on thyroid function of the rat. *Bull. environ. Contam. Toxicol.*, **7**: 19-25.

GRAHAM, S.L., DAVIS, K.J., & PERRY, C.H. (1973) Effects of one-year administration of ethylenethiourea upon thyroid function of the rat. *J. agric. food Chem.*, **21**(3): 324-329.

GRAHAM, S.L., DAVIS, K.J., HANSEN, W.H., & GRAHAM, C.H. (1975) Effects of prolonged ethylenethiourea ingestion on the thyroid of the rat. *Food Cosmet. Toxicol.*, **13**: 493-499.

GRAHAM, W.H. & BORNAK, W.E. (1973) Improved experimental technique for reverse isotope dilution method. *Anal. Chem.*, **45**: 623-624.

GRIEPENTROG, F. (1960) [Allergy studies with simple chemical substances.] *Arzneimittelforschung*, **10**: 542-544 (in German).

GRIEPENTROG, F. (1962) [Tumor-like thyroid lesions in animal studies of chronic toxicity of thiurams.] *Beitr. Pathol. Anat.*, **126**: 243-255 (in German).

GUSTAFSSON, K.H. & THOMPSON, R.A. (1981) High-pressure liquid chromatographic determination of fungicidal dithiocarbamates. *J. agric. food Chem.*, **29**: 729-732.

HAINES, L.D. & ADLER, I.L. (1973) Gas chromatographic determination of ethylenethiourea residues. *J. Assoc. Off. Anal. Chem.*, **56**(2): 333-337.

HALD, J. & JACOBSEN, E. (1948) The formation of acetaldehyde in the organism after ingestion of Antabuse (tetraethylthiuram disulfide) and alcohol. *Acta pharmacol. toxicol.*, **4**: 305-310.

HALD, J., JACOBSEN, E., & LARSEN, V. (1948) The sensitizing effect of tetraethylthiuram disulphide (Antabuse) to ethyl alcohol. *Acta pharmacol. toxicol.*, **4**: 285-296.

HEDENSTEDT, A., RANNUG, U., LAMEL, C., & WACHTMEISTER, C.A. (1979) Mutagenicity and metabolism studies on 12 thiram and dithiocarbamate compounds used as accelerators in the Swedish rubber industry. *Mutat. Res.*, **68**: 313.

HELLING, C.S., DENNISSON, D.C., & KAUFMAN, D.D. (1974) Fungicide movement in soil. *Phytopathology*, **64**(8): 1091-1100.

HOAGLAND, R.E. & FREAR, S.D. (1976) Behaviour and fate of ethylenethiourea in plants. *J. agric. food Chem.*, **24**: 129-133.

HODGE, H.C., MAYNARD, Z.A., DOWNS, W., BLANCHET, H.J., Jr, & JONES, C.K. (1952) Acute and short-term oral toxicity tests of ferric dimethyldithiocarbamate (ferbam) and zinc dimethyldithiocarbamate (ziram). *J. Am. Pharmacol. Assoc.*, **41**(12): 662-665.

HODGE, H.C., MAYNARD, E.A., DOWNS, W.L., COYE, R.D., Jr, & STEADING, L.T. (1956) Chronic oral toxicity of ferric dimethyldithiocarbamate (ferbam) and zinc dimethyldithiocarbamate (ziram). *J. Pharmacol. exp. Ther.*, **118**: 174-181.

HODGSON, J.R., CASTLES, T.R., MURRILL, E., & LEE, C.C. (1974) Distribution, excretion, and metabolism of the fungicide ferbam in rats. *Fed. Proc.*, **33**: 537.

IARC (1974) *Some antithyroid and related substances, nitrofurans and industrial chemicals*, Lyons, International Agency for Research on Cancer, pp. 45-52 (Monographs on the Evaluation of Carcinogenic Risk of Chemicals to Man, **Vol. 7**).

IARC (1976) *Some carbamates, thiocarbamates, and carbazides*, Lyons, International Agency for Research on Cancer (Monographs on the Evaluation of Carcinogenic Risk of Chemicals to Man, **Vol. 12**).

IARC (1977) *Some miscellaneous pharmaceutical substances*, Lyons, International Agency for Research on Cancer p. 243. (Monographs on the evaluation of carcinogenic risk of chemicals to humans, **Vol. 13**).

IARC (1983) *Chemicals, industrial processes and industries associated with cancer in humans*, Lyons, International Agency for Research on Cancer, pp. 128-130 (Monographs on the Evaluation of Carcinogenic Risk of Chemicals to Humans, Suppl. 4).

INNES, J.R.M., ULLAND, B.M., VALERIO, M.G., PETRUCELLI, L., FISHBEIN, L., HART, E.R., PALLOTTA, A.J., BATES, R.R., FALK, H.L., GART, J.J., KLEIN, M., MITCHELL, I., & PETERS, J. (1969) Bioassay of pesticides and industrial chemicals for tumourigenicity in mice: a preliminary note. *J. Natl Cancer Inst.*, **42**: 1101-1114.

IRPTC (1982) *Data profile on ziram*, Geneva, International Register for Potentially Toxic Chemicals.

IRPTC (1983) *Legal file - 2 vols*, Geneva, International Register for Potentially Toxic Chemicals.

IUPAC (International Union of Pure and Applied Chemistry) (1977) Ethylenethiourea. *Pure appl. Chem.*, **49**: 675-689.

IVANOVA-CHEMISHANSKA, L. (1969a) *Toxicological characteristics of dithiocarbamates zineb, maneb, and mancozeb*, Sofia, (Dissertation).

IVANOVA-CHEMISHANSKA, L. (1969b) Toxicological characteristics of mancozeb. *Hig. i. Zdraveopazvane*, **12**(5): 417-426.

IVANOVA-CHEMISHANSKA, L. & ANTOV, G. (1980) Changes in the gonads reproduction of white rats under the effect of some pesticides. *Arch. Toxicol.*, **Suppl. 4**: 459-462.

IVANOVA-CHEMISHANSKA, L., SHEYTANOV, M., & MOSHEVA-IZMIROVA, N. (1967) Changes in the functional state of the thyroid gland upon acute intoxication with certain dithiocarbamates: zineb and maneb. *Med. Toxicol.*, **20**(9): 1011-1013.

IVANOVA-CHEMISHANSKA, L., MARKOV, D.V., & DASHEV, G. (1971) Light and electron microscopic observation on rat thyroid after administration of some dithiocarbamates. *Environ. Res.*, **4**: 201-212.

IVANOVA-CHEMISHANSKA, L., KALOYANOVA, F., IZMIROVA, N., ZLATEVA, M., & VALCHEVA, V. (1972) [Experimental inhalatory toxicity and substantiation of the maximum allowable concentration for some dithiocarbamates in the air of the working area.] *Truh. Dzihot.*, **23**: 55-64 (in Russian).

IVANOVA-CHEMISHANSKA, L., VALCHEVA, V., & TAKEVA, TZ. (1973) Effect of chronic peroral poisoning with perozine (zineb) on the reproduction of white rats. *Acta med. soc.*, **1**: 107-108.

IVANOVA-CHEMISHANSKA, L., MILANOV, ST., CHEMISHANSKI, G., & DASHEV, G. (1974) Attempt of biological dose of some hypophyseal hormones in white rats after experimental oral administration of zineb. *Environ. Res.*, **8**: 160-165.

IVANOVA-CHEMISHANSKA, L., PETROVA-VERGIEVA, T., & MIRKOVA, E. (1975a) [Embryotoxic and teratogenic action of some pesticides.] *Eksp. Med. Morfol.*, **14**(1): 29-33 (in Russian).

IVANOVA-CHEMISHANSKA, L., MARKOV, D.V., MILANOV, S., STRASHIMIROV, D.D., DASHEV, G.I., & CHEMISHANKSKA, G.A. (1975b) Effects of subacute oral administration of zinc ethylene bisdithiocarbamate on the thyroid gland and adenohypophysis of the rat. *Food Cosmetol. Toxicol.*, **13**: 445-447.

IZMIROVA, N. & MARINOV, V. (1972) [Distribution and excretion of ^{35}S-ziram and metabolic products after 24 hours following oral administration of the preparation in female rats.] *Eksp. Med. Morfol.*, **11**(3): 152-156 (in Russian).

IZMIROVA, N., IZMIROVA, I., & IVANOVA, L. (1969) The effect of zineb and maneb on the isoenzymes of lactate dehydrogenase in the testes of rats. *C. R. Acad. Bulg. Sci.*, **22**(2): 225-227.

JORIS, S.J., ASPILA, K.I., & CHAKRABARTI, C.L. (1970) Decomposition of monoalkyl dithiocarbamates. *Anal. Chem.*, **42**(6): 647-651.

KAARS SIJPESTEIJN, A.K. & VONK, J.W. (1970) Microbial conversion of dithiocarbamate fungicides. *Med. Fac. Landbouwwet. Gent*, **35**(2): 799-804.

KALOYANOVA, F., IVANOVA, L., & ALEXIEV, B. (1967) The influence of large doses of maneb on the progeny of albino rats. *Med. Toxicol.*, **20**(10): 1109-1112.

KAMPMEIER, C. & HAAG, H.B. (1954) Toxicological considerations in use of dithiocarbamates. *Agric. Chem.*, **April**: 49-50, 133.

KASKEVICH, L.M., KOMAROVA, L.I., ILINA, V.I., IVANICKI, V.A., & SHILINA, V.F. (1981) [*Clinical-functional changes due to exposure to zineb*,] pp. 109-111 (in Russian).

KASLANDER, J. (1963) Formation of an *S*-glucuronide from tetraethylthiuram disulfide (Antabuse) in man. *Biochem. Biophys. Acta*, **71**: 730-732.

KATO, Y., ODANAKA, Y., TERAMOTO, S., & MATANO, O. (1976) Metabolic fate of ethylenethiourea in pregnant rats. *Bull. environ. Contam. Toxicol.*, **16**(5): 546-555.

KAUFMAN, D.D. & FLETCHER, C.L. (1973) **Title?** In: *Proceedings of the Second International Congress on Plant Pathology, Minneapolis, Minnesota* (Abstract No. 1018).

KHERA, K.S. (1973) Ethylenethiourea: teratogenicity study in rats and rabbits. *Teratology*, **7**: 243-252.

KHERA, K.S. & TRYPHONAS, L. (1977) Ethylenethiourea induced hydrocephalus: pre- and postnatal pathogenesis in offspring from rats given a single oral dose during pregnancy. *Toxicol. appl. Pharmacol.*, **42**: 85-97.

KIM, S.U. & RIZZUTO, N. (1975) Neuroaxonal degeneration induced by sodium diethyldithiocarbamate in cultures of central nervous tissue. *J. Neuropath. exp. Neurol.*, **34**: 531-541.

KING, R.R. (1977) Derivatization of ethylenethiourea with m-trifluoromethylbenzylchloride for analysis by electron capture gas chromatography. *J. agric. food Chem.*, **25**(1): 73-75.

KIRKBRIGHT, G.F. & MULLINS, F.G.P. (1984) Separation of dithiocarbamates by high-performance liquid chromatography using a micellar mobile phase. *Analyst*, **109**: 493-496.

KLIGMAN, A.M. & ROSENSWEIG, W. (1948) Studies with new fungistatic agents. I. For the treatment of superficial mycoses. *J. invest. Derm.*, **10**: 59-68.

KOIZUMI, A., SHIOJIMA, S., OMIYA, M., NAKANO, S., SATO, N., & IDEDA, M. (1979) Acute renal failure and maneb (manganous ethylene bisdithiocarbamate) exposure. *J. Am. Med. Assoc.*, **242**(23): 2583-2585.

KORHONEN, A., HEMMINKI, K., & VAINIO, H. (1982a) Embryotoxicity of industrial chemicals on the chicken embryo: thiourea derivatives. *Acta pharmacol. toxicol.*, **51**: 38-44.

KORHONEN, A., HEMMINKI, K., & VAINIO, H. (1982b) Application of the chicken embryo in testing for embryotoxicity. *Scand. J. Work environ. Health*, **8**: 63-69.

KURINNY, A.I. & KONDRATENKO, T.I. (1972) [Effect of some fungicides (dithiocarbamic acid derivatives) on chromosomes of bone marrow cells in mice.] *Tsitol. Genet.*, **6**: 225-228 (in Russian).

LABORIE, F. & LABORIE, R. (1966) Allergies aux fongicides et prophylaxie biochimique (relations entre allergie et intoxications). *Rev. Pathol. comp.*, **66**: 105-109.

LAKOMAA, E.L., SATO, S., GOLDBERG, A.M., & FRAZIER, J.M. (1982) The effect of sodium diethyldithiocarbamate treatment on copper and zinc concentrations in rat brain. *Toxicol. appl. Pharmacol.*, **65**: 286-290.

LANGE, P. & JUNG, F. (1971) [Reduction of the toxicity to the liver of carbon tetrachloride by diethyldithiocarbamate and 3-(O-tolyl)-4-(nitro)-sydnone.] *Acta biol. med. Ger.*, **27**: 425-434 (in German).

LARSSON, K.S., ARNANDER, C., CEKANOVA, E., & KJILLBERG, M. (1976) Studies of teratogenic effects of the dithiocarbamates maneb, mancozeb, and propineb. *Teratology*, **14**: 171-183.

LEE, C.C. & PETERS, P.J. (1976) Neurotoxicity and behavioural effects of thiram in rats. *Environ. Health Perspect.*, **17**: 35-42.

LEHMAN, A.J. (1951) Chemicals in food: a report of the association of food and drug officials on current developments, II, Pesticides, *Q. Bull. Fed. Drug Off. U.S.*, **15**: 122-133.

LEWERENZ, H.J., BLEYL, D., PLASS, R., & ENGST, R. (1975) *Toxicology of ETU*, Potsdam, Rehbrüke, German Democratic Republic, Central Institute for Nutrition, Academy of Science.

LIJINSKY, W., CONRAD, E., & VAN DE BOGART, R. (1972) Carcinogenic nitrosamines formed by drug/nitrite interactions. *Nature (Lond.)*, **239**: 165-167.

LOWY, R., GRIFFATON, G., BRIGANT, L., ARDONIN, B., & DUPUY, F. (1979) The dietary no-effect level of a dithiocarbamate fungicide, thiram, as evaluated from measurement data on rats. II. The various sensitivities of the various parameters. *Toxicology*, **14**: 39-53.

LOWY, R., GRIFFATON, G., & PELISIER, M.A. (1980) The dose-response relationship of rat organ weights under dietary thiram influence. *Toxicol. Lett.*, 5(Suppl. 1): 148.

LU, M.H. & STAPLES, R.E. (1978) Teratogenicity of ethylenethiourea and thyroid function in the rat. *Teratology*, **17**: 171-178.

LUDWIG, R.A. & THORN, G.D. (1962) Chemistry and mode of action of dithiocarbamate fungicides. *Adv. Pestic. Control Res.*, **3**: 219-252.

LUTZ, L.M., GLENDA, E.A., Jr, & RECKNAGEL, R.O. (1973) Protection by diethyldithiocarbamate against carbon tetrachloride lethality in rats and against carbon tetrachloride-induced lipid peroxidation *in vitro*. *Biochem. Pharmacol.*, **22**: 1729-1734.

LYMAN, W.R. (1971) The metabolic fate of Dithan M-45 (coordination product of zinc ion and manganous ethylene bisdithiocarbamate). In: Tahori, A.S., ed. *International Symposium on Pesticide Terminal Residues, Tel Aviv*, 1971, *Israel*, London, Butterworth, pp. 243-256.

LYMAN, W.R. & LACOSTE, R.J. (1974) New developments in the chemistry and fate of ethylene bisdithiocarbamate fungicides. In: *Proceedings of the 3rd International IUPAC Congress on Pesticide Chemistry, Helsinki, 3-9 July, 1974*, Stuttgart, George Thieme Publishers, pp. 67-74.

LYMAN, W.R. & LACOSTE, R.J. (1975) New developments in the chemistry and fate of ethylene bisdithiocarbamate fungicides. *Environ. Qual. Saf.*, **Suppl. 3**: 67-74.

McCANN, J. & PITCHER, F. (1973) *Fish toxicity laboratory reports Nos. 535 and 541*, Beltsville, Maryland, US Environmental Protection Agency, Animal Biology Laboratory.

MAJ, J., GRABOWSKA, M., & KWIEK, J. (1970) The effect of disulfiram, diethyldithiocarbamate, and dimethyldithiocarbamate on serotonin and 5-hydroxyindole-3-acetic acid brain levels in rats. *Biochem. Pharmacol.*, **19**: 2517-2519.

MARSHALL, W.D. (1977) Thermal decomposition of ethylene bisdithiocarbamate fungicides to ethylenethiourea in aqueous media. *J. agric. food Chem.*, **25**: 357-361.

MASSEY, R.C., KEY, P.E., & MCWEENY, D.J. (1982) Analysis of ethylenethiourea in beer by high-performance liquid chromatography. *J. Chromatogr.*, **240**: 254-256.

MATOKHNYUK, L.A. (1971) Toxicity of the fungicide maneb on inhalation. *Hyg. Sanit.*, **36**(5): 195-199.

MATTHIASCHK, G. (1973) [Influence of L-cysteine on teratogenisis caused by thiram (TMTD) in NMRI mice.] *Arch. Toxikol.*, **30**: 251-262 (in German).

MELNIKOV, N.N. & TRUNOV, P.P. (1966) Persistence of residual quantities of preparations on the basis of dithiocarbamate acid derivatives. *Chem. Agric. IV*, **10**(36): 21-24.

MINOR, J.L., RUSSEL, J.Q., & LEE, C.C. (1974) Reproduction and teratology studies with the fungicide ferbam. *Toxicol. appl. Pharmacol.*, **29**(16): 120.

MOJE, W., MUNNECKE, D.E., & RICHARDSON, L.T. (1964) Carbonyl sulfide, a volatile fungitoxicant from nabam in soil. *Nature (Lond.)*, **202**: 831-832.

NASH, R.G. (1974) Improved gas-liquid chromatographic method for determining ethylenethiourea in plants. *J. Assoc. Off. Anal. Chem.*, **57**(5): 1015-1021.

NASH, R.G. (1975) In: *Proceedings of the 170th Meeting of the American Chemical Society for Pesticides* (Abstract No. 1).

NASH, R.G. & BEALL, M.L., Jr (1980) Fate of maneb and zineb fungicides in microagroecosystem chambers. *J. agric. food Chem.*, **28**: 322-330.

NEWSOME, W.H. (1972) Determination of ethylenethiourea residues in apples. *J. agric. food Chem.*, **20**: 967-969.

NEWSOME, W.H. (1974) The excretion of ethylenethiourea by rat and guinea-pig. *Bull. environ. Contam. Toxicol.*, **11**: 174-176.

NEWSOME, W.H. (1976) Residues of four ethylene bisdithiocarbamates and their decomposition products on field-sprayed tomatoes. *J. agric. food Chem.*, **24**(5): 999-1001.

NEWSOME, W.H. (1978) A method for the determination of ethyleneurea in foods as the pentafluorobenzamide. *J. agric. food Chem.*, **26**(6): 1325-1327.

NEWSOME, W.H. & LAVER, G.W. (1973) Effect of boiling on the formation of ethylenethiourea in zineb-treated foods. *Bull. environ. Contam. Toxicol.*, **10**(3): 151-154.

ODEYEMI, O. (1980) Persistence of arasan pesticide in aquatic ecosystems in Nigeria. *Pestic. Abstr.*, **80**: 1276.

ONLEY, J.H. (1977) Gas-liquid chromatographic method for determining ethylenethiourea in potatoes, spinach, apple sauce, and milk. Collaborative study. *J. Assoc. Off. Anal. Chem.*, **60**(5): 1111-1115.

ONLEY, J.H. & YIP, G. (1971) Determination of ethylenethiourea residues in foods using thin-layer and gas chromatography. *J. Assoc. Off. Anal. Chem.*, **54**(1): 165-169.

OSKARSSON, A. (1983) Redistribution and increased brain uptake of lead in rats after treatment with diethyldithiocarbamate. *Arch. Toxicol.*, **6**: 279-284.

OSKARSSON, A. (1984) Dithiocarbamate induced redistribution and increased brain uptake of lead in rats. *Neurotoxicology*, 5(3): 283-294.

OSKARSSON, A. & LIND, B. (1985) Increased lead levels in brain after long-term treatment with lead and dithiocarbamate or thiuram derivatives in rats. *Acta pharmacol. toxicol.*, **56**: 1-6.

OTTO, S., KELLER, W., & DRESCHER, N. (1977) A new gas chromatographic determination of ethylenethiourea residues without derivatization. *J. environ. Sci. Health*, **B12**(3): 179-191.

PECKA, Z., BAULU, P., & NEWSOME, H. (1975) Preliminary survey of ethlenethiourea residues in the Canadian food supply, 1972. *Pestic. monit. J.*, **8**(4): 232-234.

PERIQUET, A. & DERACHE, R. (1976) Influence d'un régime alimentaire hypoprotéique sur la toxicité aigue d'un pesticide, le nabame. *Ann. Nutr. Aliment.*, **30**: 29-44.

PETROVA-VERGIEVA, T. (1976) On the teratogenic activity of zinc-propylene bisdithiocarbamate. *Hig. i. Zdraveopazvane*, **19**(5): 435-442.

PETROVA-VERGIEVA, T. & IVANOVA-CHEMISHANSKA, L. (1973) Assessment of the teratogenic activity of dithiocarbamate fungicides. *Food Cosmet. Toxicol.*, **11**: 239-244.

PHILLIPS, W.E.J. (1968) Changes in the nitrate and nitrite contents of fresh and processed spinach during storage. *J. agric. food Chem.*, **16**: 88-91.

PHILLIPS, W.F., GRADY, M.D., & FREUDENTHAL, R. (1977) *Effects of food processing on residues of ethylene bisdithiocarbamate fungicides and ethylenethiourea*, Washington DC, US Environmental Protection Agency (EPA No. 600/1-77-021).

PILINSKAYA, M.A. (1971) [Cytogenetic effect of the fungicide ziram in the culture of human lymphocytes *in vitro*.] *Genetika*, **7**: 138-143 (in Russian).

PLUYGERS, C.W., VONK, J.W., & THORN, G.D. (1971) Re-examination of the structure of ethylenethiuram monosulfide. *Tetrahedron Lett.*, **18**: 1317-1318.

PRZEZDZIECKI, Z., BANKOWSKA, J., KOMOROWSKA-MALEWSKA, W., & JANICKA, T. (1969) Studies of the chronic short-term toxicity of zineb, maneb, and kaptan. *Rocz. Panstw. Zakl. Hig.*, **20**: 133-140.

PRICKETT, C.S. & JOHNSTON, C.D. (1953) The *in vivo* production of carbon disulfide from tetraethylthiuram disulfide (Antabuse^{R}). *Biochem. Biophys. Acta*, **12**: 542-546.

PRINCE, J.L. (1985) Analysis of ethylenethiourea in urine by high-performance liquid chromatography. *J. agric. food Chem.*, **33**: 93-94.

RAGHU, K., MURTHY, N.B.K., KUMARASWAMY, R., RAO, R.S., & SANE, P.V. (1975) In: *Proceedings of the Joint FAO/IAEA, Division of Atomic Energy in Food and Agriculture, Vienna*, p. 137.

RASUL, A.R. & HOWELL, J.M. (1974) The toxicity of some dithiocarbamate compounds in young and adult domestic fowl. *Toxicol. appl. Pharmacol.*, **30**: 63-78.

RHODES, R.C. (1977) Studies with manganese ^{14}C-ethylene bisdithiocarbamate (^{14}C-maneb) fungicide and ^{14}C-ethylenethiourea (^{14}C-ETU) in plants, soil, and water. *J. agric. food Chem.*, **25**(3): 528-533.

RJAZANOVA, R.A. (1967) [The effects of the fungicides ziram and zineb on the generative function of test animals.] *Gig. i Sanit.*, **2**: 26-30 (in Russian).

ROLANDI, A., DE MARINIS, E., & DE CATERINA, M. (1984) Dithiocarbamate pesticides: activity of propineb in the micronucleus test in mice. *Mutat. Res.*, **135**: 193-197.

ROLL, R. (1971) [Teratological experiments with thiram (TMTD) on two mouse strains.] *Arch. Toxikol.*, **27**: 173-186 (in German).

ROSE, D., PEARSON, C.M., ZUKER, M., & ROBERTS, J.R. (1980) *Ethylenethiourea: criteria for the assessment of its effects on man*, Ottawa, National Research Council, Associate Committee on Scientific Criteria for Environmental Quality (NRCC Publication No. 18469).

ROSS, R.D. & CROSBY, D.G. (1973) Photolysis of ethylenethiourea. *J. agric. food Chem.*, **21**(3): 335-337.

ROSS HART, E. & VALERIO, M.G. (1973) Evaluation of goitrogenic activity in rats: Ethylene thiourea, practical and Dithane M-45, Communication from Litton Bionetics to Rohm and Haas Co. Cited in *Van Leeuwen et al., (1982)*

RUDDICK, J.A., WILLIAMS, D.T., HIERLIHY, L., & KHERA, K.S. (1976a) ^{14}C-ethylenethiourea: distribution, excretion, and metabolism in pregnant rats. *Teratology*, **13**: 35-39.

RUDDICK, J.A., NEWSOME, W.H., & NASH, L. (1976b) Correlation of teratogenicity and molecular structure: ethylenethiourea and related compounds. *Teratology*, **13**: 263-266.

SATO, F. & TOMIZAWA, C. (1960) Decomposition of zinc ethylene bisdithiocarbamate (zineb) applied to plants. *Bull. Natl Inst. Agric. Sci. (Tokyo) Ser. C*, **12**: 181-187.

SCHULTHEISS, E. (1957) [The therapeutic use of tetramethylthiuram disulfide.] *Dermatol. Wochenschr.*, **7**: 160-162 (in German).

SCHUPBACH, M. & HUMMLER, H. (1976) Evaluation of the mutagenicity of ethylenethiourea using bacterial and mammalian test systems. *Mutat. Res.*, **28**: 122.

SCHUPBACH, M. & HUMMLER, H. (1977) A comparative study on the mutagenicity of ethylenethiourea in bacterial and mammalian test systems. *Mutat. Res.*, **56**: 111-120.

SEIDLER, H., HARTIG, M., SCHNAAK, W., & ENGST, R. (1970) [Studies on the metabolism of certain insecticides and fungicides in the rat. II. Mean distribution and breakdown of ^{14}C-labelled maneb.] *Nahrung*, **14**(5): 363-373 (in German).

SEIFTER, J. & EHRICH, W.E. (1948) Goitrogenic compounds: pharmacological and pathological effects. *J. Pharmacol. exp. Ther.*, **92**: 303-314.

SEILER, J.P. (1973) A survey on the mutagenicity of various pesticides. *Experientia, (Basel)* **29**: 622-623.

SEILER, J.P. (1974) Ethylenethiourea (ETU): a carcinogenic and mutagenic metabolite of ethylene bisdithiocarbamate. *Mutat. Res.*, **26**: 189-191.

SEN, N.P., DONALDSON, B.A., & CHARBONNEAU, C. (1974) Formation of nitrosodimethylamine form the interaction of certain pesticides and nitrite. In: Bogorski, P., Walker, E.A., & Davis, W., ed. *N-nitroso compounds in the environment*, Lyons, International Agency for Research on Cancer, pp. 75-79 (IARC Scientific Publication No. 9).

SHERMAN, H. & ZAPP, J.A. (1966) Three generation reproduction study, Manzate D^{R} (88% maneb). Unpublished report from Haskell Laboratory submitted to the Word Health Organization by E.I. du Pont de Nemours and Co.

SHIRASU, Y., MORIYA, M., KATO, K., FURUHASHI, A., & KADA, T. (1976) Mutagenicity screening of pesticides in the microbial system. *Mutat. Res.*, **40**: 19-30.

SHIRASU, Y., MORIYA, M., KATO, K., LIENARD, F., TEZUKA, H., TERAMOTO, S., & KADA, T. (1977) Mutagenicity screening on pesticides and modification products: a basis of carcinogenicity evaluation. In: Hiatt, H.H., Watson, J.D., & Winsen, J.A., ed. *Origins of human cancer, Book A,* Cold Spring Harbor, New York, Cold Spring Harbor Laboratory, **Vol.4**, pp. 267-285.

SHORT, R.D., Jr, RUSSEL, J.Q., MINOR, J.L., & LEE, C.C. (1976) Developmental toxicity of ferric dimethyldithiocarbamate and bis(dimethylthiocarbamoyl) disulfide in rats and mice. *Toxicol. appl. Pharmacol.*, **35**: 83-94.

SHORT, R.D., MINOR, J.L., UNGER, T.M., BREEDEN, B., VAN GOETHEM, D., & LEE C.C. (1980) *Teratology of a zineb formulation.* (Results of a study performed at Midwest Research Institute for the US Environmental Protection Agency submitted to the World Health Organization) (EPA-600/1-80-017).

SMITH, D. (1976) Ethylenethiourea: a study of possible teratogenicity and thyroid carcinogenicity. *J. Soc. Occup. Med.*, **26**: 92-94.

SMITH, D.M. (1984) Ethylenethiourea: thyroid function in two groups of exposed workers. *Br. J. ind. Med.*, **41**: 362-366.

SPRECHER, D. & GRIGOROWA, R. (1967) [Maximum workplace concentration of thiuram.] *Z. gesamte Hyg.*, **13**: 241-245 (in German).

STROMME, J.H. (1965) Metabolism of disulfiram and diethyldithiocarbamate in rats with demonstration of an *in vivo* ethanol-induced inhibition of the glucuronic acid conjugation of the thiol. *Biochem. Pharmacol.*, **14**: 393-410.

STROMME, J.H. & ELDJARN, L. (1966) Distribution and chemical forms of diethyldithiocarbamate and tetraethylthiuram disulfide (disulfiram) in mice in relation to radioprotein. *Biochem. Pharmacol.*, **15**: 287-297.

SUNDERMAN, F.W., Jr & FRASER, C.B. (1983) Effects of nickel chloride and diethyl dithiocarbamate on metallothionein in rat liver and kidney. *Ann. clin. lab. Sci.*, **13**(6): 489-495.

SZYMEZYK, T. (1981) The mutagenicity of the fungicide thiram. *Mutat. Res.*, **85**: 223-224.

TAKAYAMA, S., AIHARA, K., ONODERA, T., & AKIMOTO, T. (1986) Antithyroid effects of propylthiouracil and sulfamonomethoxine in rats and monkeys. *Toxicol. appl. Pharmacol.*, **82**: 191-199.

TANNENBAUM, S.R., SINSKEY, A.J., WEISSMAN, M., & BISHOP, W. (1974) Nitrite in human saliva: its possible relationship to nitrosamine formation. *J. Natl Cancer Inst.*, **53**: 79-84.

TERAMOTO, S., MORIYA, M., KATO, K., TEZUKA, H., NAKAMURA, S., SHINGU, A., & SHIRASU, Y. (1977) Mutagenicity testing on ethylenethiourea. *Mutat. Res.*, **56**: 121-129.

TERAMOTO, S., SHINGU, A., KANEDA, M., & SAITO, R. (1978a) Teratogenicity studies with ethylenethiourea in rats, mice, and hamsters. *Congenital Anom.*, **18**: 11-17.

TERAMOTO, S., SHINGU, A., & SHIRASU, Y. (1978b) Induction of dominant lethal mutations after administration of ethylenethiourea in combination with nitrite or of *N*-nitrosoethylenethiourea in mice. *Mutat. Res.*, **56**: 335-340.

TERAMOTO, S., KANEDA, M., KATO Y. & SHIRASU Y. (1980) Failure of inducing malformations after intro-amniotic injections of ethylenethiourea in the rat. *Congenital Anom.*, **20**: 17-23.

TERAMOTO, S., KANEDA, H., AOYAMA, & SHIRASU, Y. (1981) Correlation between the molecular structure of *N*-alkylureas and N-alkylthioureas and their teratogenic properties. *Teratology*, **23**: 335-342.

TETSUMI, T., SUMI, M., TANAKA, M., & SHONO, T. (1985) Reversed phase ion interaction chromatography of sodium dialkyldithio-carbamates in the presence of tetraalkylammonium salts. *J. Chromatogr.*, **348**: 389-395.

THORN, G.D & LUDWIG, R.A. (1962) *The dithiocarbamates and related compounds*, Amsterdam, Oxford, New York, Elsevier Science Publishers, pp. 298.

TOLLENAAR, F.D. (1956) [Recent developments in the use of antioxidants in foodstuffs.] *Dtsch. Lebensm. Rundschau*, **12**: 307-313 (in German).

TRUHAUT, R., GUERINOT, F., & BOHUON, C. (1971) Mécanisme biochimique de toxicité des fongicides de la série des dithiocarbamates: L'action inhibitrice de la dopamine β-hydroxylase. *Ann. pharm. fr.*, **29**(2): 117-124.

TURNER, N.J. & CORDEN, M.E. (1963) Decomposition of sodium-N-methyldithiocarbamate in soil. *Phytopathology*, **53**: 1388-1394.

ULLAND, B.M., WEISBURGER, J.H., WEISBURGER, E.K., RICE, J.M., & CYPHER, R. (1972) Brief communication: thyroid cancer in rats from ethylenethiourea intake. *J. Natl Cancer Inst.*, **49**: 583-584.

US EPA (1977) Rebuttable presumption against registration and continued registration of pesticide products containing ethylene bisdithiocarbamates. *Fed. Reg.*, **42**(154): 40618-40627.

US EPA (1982a) *Ethylene bisdithiocarbamates: final resolution of rebuttable presumption against registration*, Washington DC, US Environmental Protection Agency, Office of Pesticides and Toxic Substances (Unpublished report).

US EPA (1982b) *Ethylene bisdithiocarbamates: notice of determination concluding the presumption against registration; notice of availability of decision document, Washington DC*, US Environmental Protection Agency (Unpublished report 82P-2851).

US EPA (1984) *Health and environmental effects: profile for ethylenethiourea (2-imidazolidinethione)*, Cincinnati, Ohio, US Environmental Protection Agency, Environmental Criteria and Assessment Office (Program Office Draft ECAO-CIN-019).

VACHKOVA-PETROVA, R. (1977) [A study on mutagenic activity of basfungine on white rats.] *Hig. i. Zdraveopazvane*, **20**(5): 422-426 (in Russian).

VACHKOVA-PETROVA, R. (1981) Mutagenicity study of endodan in rats. *Mutat. Res.*, **85**(4): 288.

VAN ESCH, G.J. (1956) [Toxicity of tetramethylthiuram disulfide and tetraethylthiuram disulfide.] *Versl. Volksgezondheid*, Vol. No. 166-169 (in Dutch).

VAN LEEUWEN, C.J. (1986) *Ecotoxicological aspects of dithiocarbamates*, Utrecht, University of Utrecht (Thesis).

VAN LEEUWEN, F.X.R., VOOGD, C.E., KNAAP, A.G.A.C., DEN TONKELAAR, E.M., & VAN DER HEIJDEN, C.A. (1982) *Toxicological evaluation of ethylenethiourea (ETU)*, Bilthoven, National Institute of Public Health (Unpublished Report No. 627915/008).

VAN LOGTEN, M.J. (1972) [*Thesis: the dithiocarbamate-alcohol reaction by the rat.*] Terborg, The Netherland, Fa. Lammers (in Dutch).

VAN STEENIS, G. & VAN LOGTEN, M.J. (1971) Neurotoxic effect of the dithiocarbamate tecoram on the chick embryo. *Toxicol. appl. Pharmacol.*, **19**: 675-686.

VEKSHTEIN, M.SH. & KHITSENKO, I.L. (1971) Ziram metabolism in warm-blooded animals. *Gig. i Sanit.*, **36**: 28-33.

VETTORAZZI, G. & VAN DEN HURK, G.W. (1984) *Pesticides reference index: JMPR 1961-84*, Geneva, World Health Organization (Unpublished report).

VOGELER, K., DREZE, Ph., RAPP, A., STEFFAN, H., & ULLEMEYER, H. (1977) Distribution and metabolism of propineb in apples and grapes, including their breakdown products propylenethiourea and ethylenethiourea in apples. *Pflanzenschutz-Nachrichten Bayer* **30**: 72-97.

VONK, J.W. & KAARS SIJPESTEIJN, A. (1970) Studies on the fate in plants of ethylene bisthiocarbamate fungicides and their decomposition products. *Ann. appl. Biol.*, **65**: 489-496.

VONK, J.W. & KAARS SIJPESTEIJN, A. (1971) Tentative identification of 2-imidazoline as a transformation product of ethylenebisdithiocarbamate fungicides. *Pestic. Biochem. Physiol.*, **1**: 163-165.

VONK, J.W. & KAARS SIJPESTEIJN, A. (1976) Formation of ethylenethiourea from 5,6-dihydro-3H-imidazo[2,1-C]-1,2,4-dithiazole-3-thione by microorganisms and reducing agents. *J. environ. Sci. Health*, **B11**: 33-47.

WATTS, R.R., STORHERR, R.W., & ONLEY, J.H. (1974) Effects of cooking on ethylene bisdithiocarbamate degradation to ethylenethiourea. *Bull. environ. Contam. Toxicol.*, **12**: 224-226.

WEDIG, J., COWAN, A., & HARTUNG, R. (1968) Some of the effects of tetramethylthiuram disulfide (TMTD) on reproduction of the bobwhite quail *(Colinus virginianus)*. *Toxicol. appl. Pharmacol.*, **12**: 293.

WEISBURGER, E.K., ULLAND, B.M., NAM, J.M., GART, J.J., & WEISBURGER, J.H. (1981) Carcinogenicity tests of certain environmental and industrial chemicals. *J. Natl Cancer Inst.*, **67**: 75-88.

WHO (1986a) *The WHO recommended classification of pesticides by hazard. Guidelines to classification 1986-1987*, Geneva, World Health Organization (Unpublished report VBC/86.1, rev. 1).

WHO (1986b) *Environmental Health Criteria 64: Carbamate pesticides: a general introduction*, Geneva, World Health Organization, pp. 137.

WHO (1988) *Environmental Health Criteria 76: Thiocarbamate pesticides: a general introduction*, Geneva, World Health Organization, pp. 49.

WORTHING, C.R. & WALKER, S.B. (1983) *The pesticide manual: a world compendium*, 7th ed., Croydon, British Crop Protection Council.

YIN-TAK WOO (1983) Carcinogenicity, mutagenicity, and teratogenicity of carbamates, thiocarbamates, and related compounds. An overview of structure-activity relationships and environmental concerns. *J. environ. Sci. Health*, **C1**(1): 97-133.

ZADOROZHNY, B.A., PETROV, B.R., OLTIRENKO, V.A., & KIRILKO, V.A. (1981) [Occupational dermatoses and their control under conditions of pesticide manufacture.] *Vestn Dermatol. Venerol.*, **6**: 48-51 (in Russian).

ZDIENICKA, M., ZIELENSKA, M., HRYMIEWICA, M., TROJANOWSKA, M., ZALEJSKA, M., & SZYMCZYK, T. (1981) The mutagenicity of the fungicide thiram. In: Kappas, A., ed. *Progress in environmental mutagenesis and carcinogenesis*, Amsterdam, Oxford, New York, Elsevier Science Publishers, **Vol. 2**, pp. 79-86.

ZORIN, P.M. (1970) [Allergic dermatitis arising as a result of contact with zineb.] *Vestn Dermatol. Venerol.*, **44**: 65-68 (in Russian).

ZUMAN, P. & ZAHRADNIK, R. (1957) [Kinetics and decomposition mechanism of dithiocarbamic acids solutions. Polarographic study.] *Z. Phys. Chem.*, **208**: 135-140 (in German).

Annex I. Names and structures of selected dithiocarbamates

Common name	Trade/other name	Chemical structure	CAS chemical name/ CAS registry number	Molecular formula	Relative molecular mass	Water solubility (25 °C)
dibam	Methylnamate	$(CH_3)_2N\text{-}\overset{\overset{S}{\|}}{C}\text{-}SNa$	sodium dimethyldi-thiocarbamate (128-04-1)	$C_3H_6NS_2Na$	143.21	
disulfiram	Antabuse	$(C_2H_5)_2N\overset{\overset{S}{\|}}{C}\text{-}SS\text{-}\overset{\overset{S}{\|}}{C}N(C_2H_5)_2$	tetraethylthiuram disulfide (29925-58-4)	$C_{10}H_{20}N_2S_4$		2 mg/litre[a]
ferbam	Fermate Fuklasin Hokmate Karbam Black Niacide	$[(CH_3)_2N\text{-}\overset{\overset{S}{\|}}{C}\text{-}S\text{-}]_3Fe$	iron, tris(dimethyl-carbamodithioato-*S,S'*)-, (14484-64-1)	$C_9H_{18}FeN_3S_6$	416.51	130 mg/litre
mancozeb	Aazimag Fore Dithane M-45 Manzate 200	$[\text{-}SCSNHCH_2CH_2NHCSSMn\text{-}]_x\ (Zn)_y$	manganese, [[1,2-ethanediylbis-[carba-modithioato]](2-)]-, in combination with [[1,2-ethanediylbis-[carbamodithioato]]-(2-)]zinc (8018-01-7)	indefinite, variable		insoluble
maneb	Amazin Blitex Dithane M-22 Manzate Martemick Mancid Tubothane	$[\text{-}SCSNHCH_2CH_2NHCSS\text{-}Mn\text{-}]_x$	manganese, [[1,2-ethanediylbis-[carba-modithioato]](2-)] (12427-38-2)	$C_4H_6MnN_2S_4$	265.29	insoluble

9

Annex I (contd).

Common name	Trade/other name	Chemical structure	CAS chemical name/ CAS registry number	Molecular formula	Relative molecular mass	Water solubility (25 °C)
metam-sodium	Carbam Masposol Sistan Trapex Vapam	CH_3NH-C(=S)-S^- Na^+	carbamodithioic acid, methyl-, sodium salt (137-42-8)	$C_2H_4NaNS_2$	129.18	722 mg/litre[b]
metiram	Zinc-metiram Polyram		ammonia complex of zineb and poly (ethylene thiuram disulfide), zineb ethylene thiuram disulfide (9006-42-2)	indefinite, variable		insoluble
nabam	Nabasan Parzate Spring-Bak	Na^+ ^-S-C(=S)-NHC_2H_4-NH-C(=S)-S^- Na^+	carbamodithioic acid, 1,2-ethanediylbis-, disodium salt (142-59-6)	$C_4H_6Na_2N_2S_4$	256.34	200 g/litre
polyram (see metiram)						
propineb	Antracol Cypromate Mezineb	[-S-C(=S)-$NHCH_2$CH(CH_3)-NH-C(=S)-S-Zn-$]_x$	zinc, [[(1-methyl-1,2-ethanediyl)-bis[carbamodithioato]](2-)]-, (12071-83-9)	$C_5H_8N_2S_4Zn$	289.9	insoluble
sulfallate	Vegadex CDEC	$(C_2H_5)_2N$-C(=S)-S-CH_2-C(Cl)=CH_2	carbamodithioic acid, diethyl-, 2-chloro-2-propenyl ester (95-06-7)	$C_8H_{14}NClS_2$	223.8	92 mg/litre

Annex I (contd).

Common name	Trade/other name	Chemical structure	CAS chemical name/ CAS registry number	Molecular formula	Relative molecular mass	Water solubility (25 °C)
thiram	Arasan Cyuram Fernasan Mercuram Normersan TMTD	$(CH_3)_2N\text{-}C(=S)\text{-}S\text{-}S\text{-}C(=S)\text{-}N(CH_3)_2$	thioperoxydicarbonic diamide, tetramethyl (137-26-8)	$C_6H_{12}N_2S_4$	240.44	30 mg/litre
zineb	Aspor-Z Carbane Dithane-Z78 Lonacol Murphane Novozir Parzate Perozine 75B Sudothane Zebenide Zelmone	$[\text{-}S\text{-}C(=S)\text{-}NHC_2H_4\text{-}NH\text{-}C(=S)\text{-}S\text{-}Zn\text{-}]_x$	zinc, [[1,2-ethane-diylbis[carbamodi-thioato]](2-)]- (12122-67-7)	$C_4H_6N_2S_4Zn$	275.73	10 mg/litre
ziram	Cuman Fuklasin Milbam Zerlate	$[(CH_3)_2N\text{-}C(=S)\text{-}S\text{-}]_2Zn$	zinc, bis(dimethyl-carbamodithioato-*S,S'*)- (137-30-4)	$C_6H_{12}N_2S_4Zn$	305.81	65 mg/litre

a At 38 °C.
b At 20 °C.

Annex II. Names and structures of degradation products of ethylene bisdithiocarbamates

Common name	Chemical structure	CAS chemical name/ CAS registry number	Molecular formula	Relative molecular mass
Ethylenethiourea (ETU)	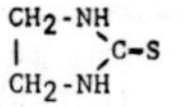	2-imidazolidinethione (96-45-7)	$C_3H_6N_2S$	102.2
DIDT	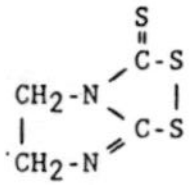	5,6-dihydro-3-*H*-imidazo[2, 1-C]1,2,4-dithiazole-3-thione, (33813-20-6)	$C_4H_4N_2S_3$	176.3
Ethylenethiuram disulfide (ETD)	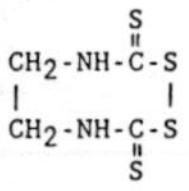	1,2,4,7-dithiadiazocine-3,8-dithione, tetrahydro (3082-38-0)	$C_4H_6N_2S_4$	210.3

Annex III. Dithiocarbamates and ETU: JMPR reviews, ADIs, Evaluation by IARC, Classification by Hazard, WHO/FAO Data Sheets, IRPTC Data Profile and Legal File[a]

Compound	Year of JMPR meeting	ADI[b] (mg/kg body weight)	Evaluation by JMPR[c]: Published in: FAO/WHO	IARC[d] Evaluation of Carcinogenicity	Availability of IRPTC[e]: Data Profile	Availability of IRPTC[e]: Legal file[g]	WHO recommended classification of pesticides by hazard[h]	WHO/FAO Data Sheets on Pesticides[f]
Ferbam	1983	0-0.02	1984a	Vol. 12 page 121	+	+	O	
	1980	0-0.02	1981b	Vol. 13 page 243				
			1981a					
	1977	0-0.02	1978b 1978a					
	1974	0-0.05 (temporary) (sum of all dithiocarbamates)	1975b 1975a					
	1970	0-0.025 (temporary) (applicable to the parent compounds only, and to the sum of all the dithiocarbamate fungicides if more than one is present)	1971b 1971a					
	1967	0-0.025 (temporary) (alone or in combination with other dimethyl-dithiocarbamates (thiram and ziram))	1968b 1968a					

Annex III (contd).

Ferbam (contd)	1965	no ADI	1965b 1965a				
	1963	no ADI	1964				
Mancozeb	1983	0-0.05[i]	1984a	+	+		0
	1980	0-0.05[i] (individually of the sum of mancozeb, maneb, and zineb)	1981b 1981a				
	1977	0-0.005 (temporary) (sum of mancozeb, maneb, and zineb)	1978b 1978a				
	1974	0-0.005 (temporary) (sum of dithiocarbamates)	1975b 1975b 1975b 1975a				
	1970	0-0.025 (temporary) (applicable to the parent compound only, and the sum of all the ethylene bisdithiocarbamate fungicides if more than one is present)	1971b 1971a				

Annex III (contd).

Compound	Year of JMPR meeting	ADI[b] (mg/kg body weight)	Evaluation by JMPR[c]: Published in: FAO/WHO	IARC[d] Evaluation of Carcinogenicity	Availability of IRPTC[e]: Data Profile	Legal file[g]	WHO recommended classification of pesticides by hazard[h]	WHO/FAO Data Sheets on Pesticides[f]
Mancozeb (contd)	1967	0-0.025 (temporary)	1968b					
		(alone or in combination with other ethylene bisdithiocarbamates (maneb and zineb), including zineb derived from nabam plus zinc sulfate)	1968a					
Maneb	1983	0-0.05[i]	1984a		+	+	O	
	1980	0-0.05	1981b					
		(individual or the sum of mancozeb, maneb, and zineb)	1981a	Vol. 12 page 137				
	1977	0-0.005 (temporary)	1978b					
		(sum of mancozeb, maneb, and zineb)	1978a					
	1974	0-0.005 (temporary)	1975b					
		(sum of all dithiocarbamates)	1975a					

Annex III (contd).

Maneb (contd)	1970	0-0.025 (temporary)	1971b			
		(applicable to the parent compound only, and to the sum of all the dithiocarbamate fungicides if more than one is present)	1971a			
	1967	0-0.025 (temporary)	1968b			
		(alone or in combination with other ethylene bisdithiocarbamates (mancozeb and zineb) including zineb derived from nabam plus zinc sulfate)	1968a			
	1965	no ADI	1965b 1965a			
	1963	no ADI	1964			
Nabam	1983	no ADI	1984a	+	+	II
	1977	no ADI	1978b 1978a			
	1974	0-0.005 (temporary)	1975b			
		(sum of all dithiocarbamates)	1975a			

Annex III (contd).

Compound	Year of JMPR meeting	ADI[b] (mg/kg body weight)	Evaluation by JMPR[c]: Published in: FAO/WHO	IARC[d] Evaluation of Carcino-genicity	Availability of IRPTC[e]: Data Profile	Legal file[g]	WHO recom-mended clas-sification of pesticides by hazard[h]	WHO/FAO Data Sheets on Pesticides[f]
Nabam (contd)	1970	0-0.025 (temporary)	1971b					
		(applicable to the parent compound only, and to the sum of all the di-thiocarbamate fung-icides if more than one is present)	1971a					
	1967	0-0.025 (temporary)	1968b					
		(as nabam alone or in combination with other ethylene bisdithiocarbamates (mancozeb, maneb, and zineb) including zineb derived from nabam plus zinc sulfate)	1968a					
	1965	no ADI	1965b 1965a					
	1963	no ADI	1964					
Propineb	1985	ADI withdrawn	1986b				0	
	1984	0-0.005 (temporary)	1985b					
	1983	0-0.005 (temporary)	1984b					

Annex III (contd).

Propineb (contd)	1980	0-0.005 (temporary)	1981b				
	1977	0-0.005	1978b				
Thiram	1983	0-0.005 (temporary)	1984a	Vol. 12 page 225	+	+	III
	1980	0-0.005 (temporary)	1981b 1981a				
	1977	0-0.005 (temporary)	1978a				
	1974	0-0.005 (temporary)	1975b				
		(sum of all dithiocarbamates)	1975a				
	1970	0-0.025 (temporary)	1971b				
		(applicable to the parent compound only, and to the sum of all the dithiocarbamate fungicides if more than one is present)	1971a				
	1967	0-0.025 (temporary)	1968b				
		(alone or in combination with other dimethyl dithiocarbamates (ferbam and ziram))	1968a				

Annex III (contd).

Compound	Year of JMPR meeting	ADI[b] (mg/kg body weight)	Evaluation by JMPR[c]: Published in: FAO/WHO	IARC[d] Evaluation of Carcinogenicity	Availability of IRPTC[e]: Data Profile	Legal file[g]	WHO recommended classification of pesticides by hazard[h]	WHO/FAO Data Sheets on Pesticides[f]
Thiram (contd)	1965	0-0.025	1965b 1965a					
	1963	0-0.025	1964					
Zineb	1983	0-0.05[i]	1984a	Vol. 12 page 245	+	+	O	
	1980	0-0.05 (individually or the sum of mancozeb, maneb, and zineb)	1981b 1981a					
	1977	0-0.005 (temporary)	1978b					
		(sum of mancozeb, maneb, and zineb)	1978a					
	1974	0-0.005 (temporary)	1975b					
		(sum of all dithiocarbamates)	1975a					
	1970	0-0.025 (temporary)	1971b					
		(applicable to the parent compound only, and to the sum of all the dithiocarbamate fungicides if more than one is present)	1971a					
	1967	0-0.025 (temporary)	1968b					

Zineb (contd)	1967	(alone or in combination with other ethylene bisdithiocarbamates (mancozeb and maneb) including zineb derived from nabam plus zinc sulfate)	1968a					
	1965	no ADI	1965b 1965a					
	1963	no ADI	1964					
Ziram	1983	0-0.02	1984a	Vol. 12 page 259	+	+	III	No. 73 (in preparation)
	1980	0-0.02	1981b 1981a					
	1977	0-0.02	1978b 1978a					
	1974	0-0.005 (temporary)	1975b					
		(sum of all dithiocarbamates)	1975a					
	1970	0-0.025 (temporary)	1971b					
		(applicable to the parent compound only, and to the sum of all the dithiocarbamate fungicides if more than one is present)	1971a					
	1967	0-0.025 (temporary)	1968b					
		(alone or in combination with other dimethyl dithiocarbamates (ferbam and thiram))	1968a					
	1965 1965a	no ADI	1965b					
	1963	no ADI	1964					

Annex III (contd).

Compound	Year of JMPR meeting	ADI[b] (mg/kg body weight)	Evaluation by JMPR[c]: Published in: FAO/WHO	IARC[d] Evaluation of Carcinogenicity	Availability of IRPTC[e]: Data Profile	Legal file[g]	WHO recommended classification of pesticides by hazard[h]	WHO/FAO Data Sheets on Pesticides[f]
ETU (see dithiocarbamates)	1980	0.002	1981b	Vol. 7, p. 45				
	1974	-	1975b	Suppl. 4, p. 128				
PTU (see propineb)	1985	no ADI (withdrawn)	1986a					

a Adapted from: Vettorazzi & van den Hurk (1984).
b ADI = acceptable daily intake.
c JMPR = Joint Meeting on Pesticide Residues (FAO/WHO).
d IARC = International Agency for Research on Cancer (WHO, Lyons, France).
e IRPTC = International Register for Potentially Toxic Chemicals (UNEP, Geneva).
f WHO/FAO Data Sheets on Pesticides with number and year of appearance.
g From: IRPTC (1983).
h From: WHO (1986a).
i Not more than 0.002 mg/kg body weight may be present as ETU.

The hazard referred to in this Classification is the acute risk for health (that is, the risk of single or multiple exposures over a relatively short period of time) that might be encountered accidentally by an person handling the product in accordance with the directions for handling by the manufacturer or in accordance with the rules laid down for storage and transportation by competent international bodies.

Classification relates to the technical material, and not to the formulated product:

Class		LD_{50} for the rat (mg/kg body weight): Oral		Dermal	
		Solids	Liquids	Solids	Liquids
1A	Extremely hazardous	5 or less	20 or less	10 or less	40 or less
1B	Highly hazardous	5 - 50	20 - 200	10 - 100	40 - 400
II	Moderately hazardous	50 - 500	200 - 2000	100 - 1000	400 - 4000
III	Slightly hazardous	over 500	over 2000	over 1000	over 4000
O	Unlikely to present acute hazard in normal use				

Other titles available in the ENVIRONMENTAL HEALTH CRITERIA series (continued):

44. Mirex
45. Camphechlor
46. Guidelines for the Study of Genetic Effects in Human Populations
47. Summary Report on the Evaluation of Short-Term Tests for Carcinogens (Collaborative Study on *In Vitro* Tests)
48. Dimethyl Sulfate
49. Acrylamide
50. Trichloroethylene
51. Guide to Short-Term Tests for Detecting Mutagenic and Carcinogenic Chemicals
52. Toluene
53. Asbestos and Other Natural Mineral Fibres
54. Ammonia
55. Ethylene Oxide
56. Propylene Oxide
57. Principles of Toxicokinetic Studies
58. Selenium
59. Principles for Evaluating Health Risks from Chemicals During Infancy and Early Childhood: The Need for a Special Approach
60. Principles and Methods for the Assessment of Neurotoxicity Associated With Exposure to Chemicals
61. Chromium
62. 1,2-Dichloroethane
63. Organophosphorus Insecticides — A General Introduction
64. Carbamate Pesticides — A General Introduction
65. Butanols — Four Isomers
66. Kelevan
67. Tetradifon
68. Hydrazine
69. Magnetic Fields
70. Principles for the Safety Assessment of Food Additives and Contaminants in Food
71. Pentachlorophenol
72. Principles of Studies on Diseases of Suspected Chemical Etiology and Their Prevention
73. Phosphine and Selected Metal Phosphides
74. Diaminotoluenes
75. Toluene Diisocyanates
76. Thiocarbamate Pesticides — A General Introduction
77. Man-made Mineral Fibres